Research Methodolc
Approaches and Techni

Research Methodology
Approaches and Techniques

Dr. Bhawna Garg

AMIGA PRESS. INC
Delhi -110051

AMIGA
PRESS INC.
AMIGA PRESS INC.
Registered Office:
19 DBS Building, Ansari Road, Darya Ganj, Delhi. 110002, India
Phone: +91-11-23283333 Email: amigapress@gmail.com

This edition is published by Amiga Press.Inc

ISBN : 978-93-84533-01-4

First Published 2018

Reprint 2024

Printed and bound in India by Replika Press Pvt. Ltd.

Preface

Studying the subject of research methodology is the common to almost all the postgraduate students those who want to undergo the academic research or field work. Research is the part of the business which is concentrating the profit maximization through innovative product, market strategies and systematic administration. For all the above mentioned activities, research is the backbone and inducing mechanism.

The purpose of organizational behaviour is to gain a greater understanding of those factors that influence individual and group dynamics in an organizational setting so that individuals and the groups and organizations to which they belong may become more efficient and effective. The field also includes the analysis of organizational factors that may have an influence upon individual and group behaviour. Much of organizational behaviour research is ultimately aimed at providing human resource management professionals with the information and tools they need to select, train, and retain employees in a fashion that yields maximum benefit for the individual employee as well as for the organization.

Organizational behaviour is a relatively new, interdisciplinary field of study. Although it draws most heavily from the psychological and sociological sciences, it also looks to other scientific fields of study for insights. One of the main reasons for this interdisciplinary approach is because the field of organizational behaviour involves multiple levels of analysis, which are necessary to understand behaviour within organizations because people do not act in isolation. That is, workers influence their environment and are also influenced by their environment.

In responding to the challenge of globalisation, businesses entities are organised differently compared to the past. Changes have been made to the supply chain in the form of new development such as company organisational structure, just-in time delivery, warehousing and logistics operation. Before, businesses wanted to own all the activities encompassed in their supply chain, but now they only control the strategic core of their supply chain activities and depend more on external business partners.

Strategic planning is an organization's process of defining its strategy, or direction, and making decisions on allocating its resources to pursue this strategy, including its capital and people. Various business analysis techniques can be used in strategic planning, including SWOT analysis and PEST analysis. The outcome is normally a strategic plan which is used as guidance to define functional and divisional plans, including Technology, Marketing, etc.

This is necessary to avoid oversight or inattention by the management. An informal plan is one which is not reduced to writing but is kept only in the mind of the manager. If the number of actions to be taken is less and the actions have to be taken in a short period informal planning may be usefully adopted. Plans may be short-term or long-term depending upon the period which is kept in view while formulating a plan.

Similarly, there may be single-use or standing plans. A single-use plan is one which sets a course of action for a specific situation and is used up once the particular goal is reached, whereas the standing plan is one which is designed to be used over and over again.

Planning is to be done at all levels of organisation, but the lower-level management tends to derive its own plans from the plan of higher-level management. However, higher-level management spends more time and energy in planning than lower-level management. Higher-level management keeps broader objective of the company as a whole and takes a long-run view whereas others keep in mind only immediate objective and sectional approach in planning.

Research methodology for management is the book which is purely based on the university syllabus. The presentation of the language and explanation of the concept will be based in the student's capacity.

—Editor

Contents

1

Research Methodology

The system of collecting data for research projects is known as research methodology. The data may be collected for either theoretical or practical research for example management research may be strategically conceptualized along with operational planning methods and change management. Some important factors in research methodology include validity of research data, Ethics and the reliability of measures most of your work is finished by the time you finish the analysis of your data.

Formulating of research questions along with sampling weather probable or non probable is followed by measurement that includes surveys and scaling.

QUANTITATIVE RESEARCH

In the social sciences, quantitative research refers to the systematic empirical investigation of quantitative properties and phenomena and their relationships. The objective of quantitative research is to develop and employ mathematical models, theories and/or hypotheses pertaining to phenomena. The process of measurement is central to quantitative research because it provides the fundamental connection between empirical observation and mathematical expression of quantitative relationships. Quantitative research is used widely in social sciences such as psychology, sociology, anthropology, and political science. Research in mathematical sciences such as physics is also 'quantitative' by definition, though this use of the term differs in context.

In the social sciences, the term relates to empirical methods, originating in both philosophical positivism and the history of statistics, which contrast qualitative research methods. Qualitative methods produce information only on the particular cases studied, and any more general conclusions are only hypotheses. Quantitative methods can be used to verify, which of such hypotheses are true. A comprehensive analysis of 1274 articles published in the top two American sociology journals between 1935 and 2005 found that roughly two thirds of these objects used quantitative methods

Quantitative research is generally made using scientific methods, which can include:

- Collection of empirical data
- Evaluation of results
- Experimental control and manipulation of variables
- Modeling and analysis of data
- The development of instruments and methods for measurement
- The generation of models, theories and hypotheses

In the social sciences particularly, quantitative research is often contrasted with qualitative research which is the examination, analysis and interpretation of observations for the purpose of discovering underlying meanings and patterns of relationships, including classifications of types of phenomena and entities, in a manner that does not involve mathematical models.

Approaches to quantitative psychology were first modelled on quantitative approaches in the physical sciences by Gustav Fechner in his work on psychophysics, which built on the work of Ernst Heinrich Weber. Although a distinction is commonly drawn between qualitative and quantitative aspects of scientific investigation, it has been argued that the two go hand in hand. For example, based on analysis of the history of science, Kuhn concludes that "large amounts of qualitative work have usually been prerequisite to fruitful quantification in the physical sciences".

Qualitative research is often used to gain a general sense of phenomena and to form theories that can be tested using further quantitative research. For instance, in the social sciences qualitative research methods are often used to gain better understanding of such things as intentionality and meaning. Although quantitative investigation of the world has existed since people first began to record events or objects that had been counted, the modern idea of quantitative processes have their roots in Auguste Comte's positivist framework. Statistics is the most widely used branch of mathematics in quantitative research outside of the physical sciences, and also finds applications within the physical sciences, such as in statistical mechanics. Statistical methods are used extensively within fields such as economics, social sciences and biology. Quantitative research using statistical methods starts with the collection of data, based on the hypothesis or theory. Usually a big sample of data is collected - this would require verification, validation and recording before the analysis can take place.

Software packages such as SPSS and R are typically used for this purpose. Causal relationships are studied by manipulating factors thought to influence the phenomena of interest while controlling other variables relevant to the experimental outcomes.

In the field of health, for example, researchers might measure and study the relationship between dietary intake and measurable physiological effects such as weight loss, controlling for other key variables such as exercise. Quantitatively based opinion surveys are widely used in the media, with statistics such as the proportion of respondents in favour of a position commonly reported. In opinion surveys, respondents are asked a set of structured questions and their responses are tabulated.

In the field of climate science, researchers compile and compare statistics such as temperature or atmospheric concentrations of carbon dioxide. Empirical relationships and associations are also frequently studied by using some form of General linear model, non-linear model, or by using factor analysis.

A fundamental principle in quantitative research is that correlation does not imply causation. This principle follows from the fact that it is always possible a spurious relationship exists for variables between which covariance is found in some degree. Associations may be examined between any combination of continuous and categorical variables using methods of statistics. Views regarding the role of measurement in quantitative research are somewhat divergent. Measurement is often regarded as being only a means by which observations are expressed numerically in order to investigate causal relations or associations.

However, it has been argued that measurement often plays a more important role in quantitative research. For example, Kuhn argued that within quantitative research, the results that are shown can prove to be strange. This is because accepting a theory based on results of quantitative data could prove to be a natural phenomenon. He argued that such abnormalities are interesting when done during the process of obtaining data:

- When measurement departs from theory, it is likely to yield mere numbers, and their very neutrality makes them particularly sterile as a source of remedial suggestions. But numbers register the departure from theory with an authority and finesse that no qualitative technique can duplicate, and that departure is often enough to start a search.

In classical physics, the theory and definitions which underpin measurement are generally deterministic in nature. In contrast, probabilistic measurement models known as the Rasch model and Item response theory models are generally employed in the social sciences. Psychometrics is the field of study concerned with the theory and technique for measuring social and psychological attributes and phenomena. This field is central to much quantitative research that is undertaken within the social sciences. Quantitative research may involve the use of proxies as stand-ins for other quantities that cannot be directly measured. Tree-ring width, for example, is considered a reliable proxy of ambient environmental conditions such as the warmth of growing seasons or amount of rainfall. Although scientists cannot directly measure the temperature of past years, tree-ring width and other climate proxies have been used to provide a semi-quantitative record of average temperature in the Northern Hemisphere back to 1000 A.D. When used in this way, the proxy record only reconstructs a certain amount of the variance of the original record. The proxy may be calibrated to determine how much variation is captured, including whether both short and long term variation is revealed.

In the case of tree-ring width, different species in different places may show more or less sensitivity to, say, rainfall or temperature: when reconstructing a temperature record there is considerable skill in selecting proxies that are well correlated with the desired variable. Quantitative methods are research techniques that are used to gather quantitative data — information dealing with numbers and anything that is measurable. Statistics, tables and graphs, are often used to present the results of these methods. They are therefore to be distinguished from qualitative methods. In most physical and biological sciences, the use of either quantitative or qualitative methods is uncontroversial, and each is used when appropriate. In the social sciences, particularly in sociology, social anthropology and psychology, the use of one or other type of method has become a matter of controversy and even ideology, with particular schools of thought within each discipline favouring one type of method and pouring scorn on to the other.

Advocates of quantitative methods argue that only by using such methods can the social sciences become truly scientific; advocates of qualitative methods argue that quantitative methods tend to obscure the reality of the social phenomena under study because they underestimate or neglect the non-measurable factors, which may be the most important. The modern tendency is to use eclectic approaches. Quantitative methods might be used with a global qualitative frame. Qualitative methods might be used to understand the meaning of the numbers produced by quantitative methods. Using quantitative methods, it is possible to give precise and testable expression to qualitative ideas. This combination of quantitative and qualitative data gathering is often referred to as mixed-methods research.

METHODOLOGY OF THE RESEARCH PROJECT

In 1998, our research team undertook a qualitative study on 24 pedophiles in treatment at a specialized clinic for sex offenders. This study was to be my doctoral thesis. The initial goals of this study were to understand (1) What motivates pedophile sex offenders to enter and remain in a specialized treatment program, and (2) How they experience the treatment on a daily basis. Nondirective semi-structured interviews (Kandel, 1972; Legras, 1971) were used to gather the subjects' thoughts about their experience of the treatment as well as what they considered to have motivated them to enter and remain in the treatment program. The interviews were analysed using diverse qualitative methods, including text-based comparative analysis (Maykut & Morehouse, 1994), Multiple-case depth analysis (Schneider, 1999), and psychoanalytically-based qualitative analysis.

In order to examine specific questions of interest, Plan Analysis (Caspar, 1995, 1997) was also used on a selected number of subjects. Plan Analysis was done by independent judges based on qualitative theme analysis. It involves gathering clinically relevant information about an individual's behaviour and experience through careful observation and synthesizing it into a meaningful whole. As Caspar writes (1997), "the fundamental question that guides Plan Analysis is as follows: which purpose, conscious or unconscious, underlies an individual's behaviors and experiences?" The different components of a Plan Analysis are then interrelated and hierarchically organized. They include a patient's goals or desires but also the means by which these goals are attained. Some obvious advantages of Plan Analysis are that it takes into consideration many different theories, it is idiosyncratic and robust and it is pragmatic and relies on concrete methodology.

Finally, after using the qualitative methods mentioned above, some specific pre-established hypotheses derived from the literature on sex abusers were tested using quantitative methods on selected cases. The first method used was Perry's Wish and Fear List (1996). The Wish & Fear List is based on the notion that behaviour is motivated. These motives can be both biologically and psychologically determined in varying degrees and consist, in their simplest form, of wishes and fears. According to Perry (1996), "a dynamic *Wish* differs from a simple desire by playing a causal role in a variety of behaviors, fantasies and experiences". A *Fear* can be defined as "a negative belief, expectation, or aversive experience which the subject wishes to avoid". The method developed by Perry encompasses notions of Erikson's developmental stages and of self-psychology and reintroduces an object-relations perspective via an interest for motives. Such a perspective, although avowedly psychodynamic, has the advantage of being free of strict psychoanalytic jargon and of not violating any basic propositions from learning theory.

Finally, the last instrument used was the Core Conflictual Relationship Theme (CCRT) (Luborsky & Crits-Christoph, 1990). Amongst others things, the CCRT identifies three components: a subject's wishes (W), the responses from others (RO), and the responses of self (RS).

There is much evidence that these components describe core repetitive relationship patterns and share many characteristics with Freud's transference theory. The CCRT was used by one independent judge in order to examine our subjects' relationship patterns.

A research topic like sex abuse-even more so when it concerns children and pedophilic behaviour- presents many major difficulties. The first one a researcher is confronted to is the question of definition of sex abuse and, more specifically, of sexual abuse of children.

Despite the spectacular increase of research in this field, no consensus has yet been reached. Between the writings of Krafft-Ebing (1912) and the more recent definition of pedophilia in the DSM-IV (APA, 1994) can be found a multitude of definitions and concepts often revealing strong legal and moral implications. Some of these definitions, still in use today, include an individual masturbating while thinking or fantasizing about children, a child being victim of an exhibitionist or a more clearly defined sexual contact between a child and an adult.

In that sense, some definitions are still the result of strong prejudice or personal conflicts. It is then not surprising to see that sexology has often been infiltrated by demonology and that many examples mentioned in the literature clearly are limited to the most provocative cases. This has often led to certain exaggerations such as considering all sexual activity of or between children as abusive (Finkelhor, 1979).

The problem of studying sexual abuse of children and pedophilic behaviour is even greater when one considers the reactions a research project may bring up. Since many studies are often used in court in order to obtain either a guilty or a non-guilty verdict, some researchers have been thought of in the public opinion as being pro-pedophile.

The least that can be said is that the study of sexual abuse is an emotionally charged one. As my study was on the motivation for treatment of pedophilic sexual abusers, I could not help but confront these questions as well as what brought me to give attention to this topic. But my main concern (and what I wish to emphasize here) was what could be understood from interviewing pedophilic subjects. With all of this in mind and as I was in a psychoanalytic cure at the time, I proceeded to bring up these questions with my analyst, who later on acted as "third party".

Subjectivity in Understanding the Subjects

My first subject was a man in his sixties. Prior his arrest, he had held an important and very well considered professional position. He was serving a long-term federal sentence after being condemned for sexually abusing young boys over a period of many years.

He showed up at the interview very well dressed and looking confident. He answered all questions without hesitation, giving much detail in a way a professor would lecture his students. After 70 minutes of interviewing, the meeting was interrupted by a guard asking that the subject return in his cell for inspection. I then proceeded to note down my subjective impressions in a notebook I reserved for this use. These close to two pages of notes included the following:

"(Subject) seems very educated and definitely very articulate. His strong accent gives him a sort of aristocratic twist. It felt as if he was actually lecturing me or teaching me something that I should know. (I) felt like a kid who's begging for information and desperately waiting for his next word. Seemed like I was learning something !!! (…) The contact was good, maybe even a little too good? Something fascinating about him?, almost father or grandfather like (…) What is it? (…) At the same time, it felt intrusive, as if I should better watch out and not get skrewed. Don't know what to make of it (…) Very confusing, ambivalent."

The complete notes were later on brought in an analytic session and a total of close to four sessions were actually used to try to understand what felt during this interview. The most striking aspect was felt like twenty or twenty-five years younger, almost like a little boy.

It was felt charmed and fascinated by this person as if could learn much from him. Through analysis, anyhow managed to establish clear relations between this and my own history, more precisely with my father. In that sense, it seemed as if was longing for something that was missing and that this subject could actually give me, that is: a feeling of being an "adult, mature man".

As such, during the interview, there were *fantasies* of establishing contact with him once his imprisonment would be over, hoping that what was missed during the interview could be obtained later through some kind of contact. While having these feelings, it was also felt as if it should better make sure one don't get "skrewed". This last term brought up much anxiety, as it could easily be related to sexual abuse. And somehow, it did feel like rape. It then became clear that it was felt ambivalent towards the subject, both fearing and wanting him to give me the feeling of being an "adult man". Something the subject lived somehow resonated in me, while at the same time not being completely mine. This led to the idea that, although the subject and was shared certain common fantasies, differences could be found. It so appeared that through different processes, namely projective identification and identification to the aggressor, the subject had led me on a *playground* where his rules prevailed and where he could actually bring me to feel like he might have felt while maintaining himself in a superior-fatherly- position. Spontaneously, this leads to many hypotheses regarding the subject's functioning:

1. He might have missed having a strong-if any- father figure capable of leading him to adult life;
2. He may have struggled with an authoritarian maternal figure; and
3. He may feel like a young boy and not entirely like an adult man and these feelings are intolerable for him.

It was later realized, many psychoanalytic theories have been developed around these ideas and identification to the aggressor is often a core component of them. According to Laplanche and Pontalis (1967), identification to the aggressor involves the subject identifying to his aggressor by making himself the perpetuator or making his certain characteristics of the aggressor. As such, the pedophilic act involves mastery by leaving the victim position to become the aggressor. More recent theories (Balier, 1993, 1996) also suggest that the pedophile is in many ways "his mother's narcissistic doll". Indirectly, this suggestion raises the question as to what place-if any- the father had in the child's life. It also suggests that the pedophile struggles with symbiotic issues with the mother and lacks positive identification with the father. Through analysis with a "third party", these hypotheses seemed applicable. But the question still remained as to how true or significant they were.

Verifying these hypotheses As quantitative measures had been used in order to examine other aspects of the research question, quantitative results were readily available for the subject presented above. Hence, it was interesting to triangulate hypotheses resulting from the examination of my subjectivity with other methods. Using Caspar's Plan Analysis (1997), it appeared that some of the most central themes brought up by this subject were "a need to have a sense of mastery", "a fear of being dominated", and "a wish to put the therapists to test by confronting them in order to make sure they are strong, authoritarian and reliable "father" figures". The Wish and Fear List was used by independent judges and suggested that the subject's most prevalent motives were: a wish to be autonomous, a wish to control and dominate others, a wish to have and be a role model or a mentor, a fear of being dominated, and a fear of physical harm.

Overall, most motives mentioned by the subject were related to what Erikson (1985) called autonomy versus shame and doubt issues. Finally, the CCRT indicated that the subject was most often motivated to have control over others, to be independent, and to be his own person. Others most often reacted to these wishes by being controlling (in the case of women) or being distant- feeling controlled (in the case of men). The subject then was left feeling helpless, incompetent and disappointed.

What is interesting about these methods is that they all seem to underline and empirically support different aspects of what had been *subjectively* felt during the interview. Hence, examining my subjectivity provided-in itself- a good indication of the subject's functioning. But as it will be described next, examining one's subjectivity can also be important when the time comes to analyse the data at hand as other factors also have an influence on the research process.

Subjectivity in the Research Mentor

As it was intensive training in quantitative research and still worked in that field, chosen to do a Ph.D. where one could learn new and different research methods. This led me to a psychoanalysis professor who had been doing qualitative work for many years. Very early in the research project, a deep abyss could already be seen between the way this professor saw therapeutic work and the way patients were treated at the penitentiary.

All pedophilic patients there had the opportunity to start a treatment program in a specialized clinic associated to the penitentiary. This treatment, as most programs for sexual abusers, was cognitive-behavioural oriented and included the use of aversive conditioning using ammonia or, occasionally, mild self-inflicted electroshocks. Although such treatments have been proven most effective with sexual abusers and while psychoanalytically orientated therapies had largely failed in this matter, my professor was somewhat shocked and offended when it was first described the treatment to him.

In the months that followed, it was ended up being a moderator for both parties, that is the cognitive-behavioural and the old fashioned psychoanalytic models. Clearly enough, as my professor was the one who not only rated my work but had the final word on the thesis project, it was made clear efforts when analyzing the data in order to demonstrate his impressions were right: "people just cannot be satisfied with such a rigid treatment that gives them so little room!". For example, it would be would over emphasize the fact that the subjects had little chance to talk (which is a central aspect of psychoanalytic work) despite their desire-and capacity- to do so. In that sense, when one first analysed the interviews, took special care to show that the subjects had a strong wish to talk, while looking for clues that they were not satisfied with the present treatment. For these reasons, it would be over emphazise a subject's strengths and capacities in order to demonstrate that he could be amenable to psychoanalytic therapy but was deprived of it.

As work with my research mentor eventually became difficult and even counter-productive, one became more and more frustrated and then switched positions. In analyzing data, it would be given more attention to clues indicating that the subject was *not* amenable to psychoanalytic therapy. At the same time, it would be over emphasized the "cruelty" (as he put it) of the cognitive-behavioural treatment. It became clear that my efforts, although not entirely conscious at the time, aimed at contradicting and offending him. This confrontation was motivated by technical as well as emotional motives. For instance, although it quickly became clear, through analysis, that it was deliberately confronted with research mentor in order to contradict him as a result of being frustrated for having been placed in a moderator position, there also appeared to be other reasons for doing so. One of these reasons seems of particular importance and was the result of certain "similarities" between the relation had with my supervisor and the one my first subject had with his father (as seen in theme analysis): neither one would "put out", that is give what they were supposed to give (e.g., support and input). Because of this, it was identified easily with the subject and used his (most shocking) words to confront my supervisor, as the subject had done himself with his father. The only way to overcome these difficulties was to analyse them with the help of the "third party".

QUANTITATIVE METHODS

- *Unit of Analysis*: The most elementary part of what is being studied or observed. Some examples include individuals, households, court cases, countries, states, firms, industries, etc.
- *Variables*: Concepts, characteristics, or properties that can vary, or change, from one unit of analysis to another. Please note that all variables must vary, if there is no variation among the different cases then it is not a variable. Some examples of variables include gender, social class, education, age, level of public enforcement, type of bankruptcy, etc.
  - *Dependent Variable*: Variables whose change the researcher wishes to explain
  - *Independent Variable*: Variables that help explain the change in the dependent variable
- *Hypothesis*: An empirical statement which seeks to test the relationship between at least two variables. For instance, As levels of public enforcement increases, levels of stock development also increases. This hypothesis has two variables:
  - Public enforcement{independent variable, and
  - Stock development{dependent variable.
- Levels of Measuring Variables
  - *Nomial*: A nominal variable has qualitative categories that can- not be ranked in a meaningful way in terms of degree or magnitude. Examples of nominal variables include race, type of bankruptcy, type of corporation, name. All of these variables have qualitative categories that cannot be ordered in terms of magnitude or degree. This is the least powerful type of variable.

- *Ordinal*: An ordinal variable has qualitative categories that are ordered in terms of degree or magnitude. Examples of a nominal variable include class or dgree obtained. The variable DEGREE OBTAINED may include the following categories: None, High School Diploma, College/University Degree, Masters, Advanced Degree (JD/PHD/MD). All of these categories are qualitative and are ordered in terms of the amount of education each individual has completed.
- *Interval/Ratio*: An interval variable has quantitative values (or numbers). Some examples of interval variables include AGE (in years), number of shares outstanding, and Amount in DEBT (in dollars). For all of these variables the response is going to be a number or value. This is most powerful type of variable because you can do the most with it statistically. Note that if a variable has qualitative categories that ARE ordered and there are numerical values assigned to each category which are also ordered, we can treat this variable like an interval level variable. An example would be questionnaire that asks respondents about their feelings towards President Obama's handling of the economy on a scale of 1 to 5 where (1=very bad job, 2=bad job, 3=neither bad nor good, 4=good job, and 5=very good job). The respondents are asked to choose a category that is ordered, but since it has ordered numbers attached to the categories, we can treat it as an interval level variable with some restrictions.
- *Dichotomous/Dummy*: A dichotomous variable is a variable with two (and only two) categories. These categories can be qualitative or quantitative values.

THE QUALITY OF QUALITATIVE RESEARCH

Probably many qualitative researchers have been criticised by critics who have claimed that their research it is too subjective, or the number of cases is too small, or that mere talking is never a scientific method, and so on. Qualitative inquiry, moreover, has a short history in Japanese social work research and therefore Japanese social work researchers may need to strongly defend the validity of their qualitative research. However, they should do so in the confidence that qualitative research is no less rigorous than more traditional forms of inquiry.

Although "validity" and "reliability" of qualitative research is discussed by many researchers, the most often quoted concept of the problem of establishing validity is probably the notion of trustworthiness that was developed mainly by Lincoln and Guba (1985). The notion of trustworthiness has four elements: credibility, transferability, dependability, and confirmability. These are analogous to "internal validity", "external validity", "reliability", and "objectivity" in conventional criteria. Because their philosophical base is constructivism, these criteria are not applied to all the qualitative research methods. We believe, however, that these criteria are useful for novice researchers to understand the validity and reliability issues of qualitative research and so we will describe them and some of the techniques related to them.

Credibility, Prolonged Engagement and Peer Debriefing

Credibility is analogous to "internal validity" in conventional criteria. It relates to how the reconstruction of the researchers fits the realities and views the participants express in the process of the inquiry. To establish credibility, researchers use a variety of techniques, but because we have limited space, we will discuss only the following. One is "prolonged engagement." That is "the investment of sufficient time to achieve certain purposes; learning the 'culture [of the participants],' testing for misinformation introduced by distortions either of the self or of the respondents, and building trust [with the participants]". Another is "peer debriefing." This is the process of "allowing a peer who is a professional outside the context and who has some general understanding of the study to analyse materials, test working hypotheses and emerging designs, and listen to the researcher's ideas and concerns".

Transferability and Thick Descriptions

Transferability refers to the possibility that what was found in one context by a piece of qualitative research is applicable to another context. As Lincoln and Guba (1985) point out: If there is to be transferability, the burden of proof lies less with the original investigator than with the person seeking to make an application elsewhere. The original inquirer cannot know the sites to which transferability might be sought, but the appliers can and do.... The responsibility of the original investigator ends in providing sufficient descriptive data to make such similarity judgements possible.

Providing sufficient descriptive data is often called "thick description". For example, if you conduct qualitative research on a day centre for people with mental disabilities, you should describe the context so thickly that users of your research report can judge whether the findings are applicable to their own settings.

Dependability, Confirmability and Auditing

Dependability is the qualitative researcher's equivalent of the conventional term "reliability", which is equal to replicability. In quantitative research, reliability means that the same tests should produce the same results. For qualitative researchers, this kind of replicability is impossible to realise because the research design is so flexible and the research findings are produced by constantly changing interactions between researchers and participants. Therefore, as Guba and Lincoln (1989) states, Far form being threats to dependability, such changes and shifts are hallmarks of a maturing-and successful-inquiry. But such changes and shifts need to be both tracked and trackable. This idea leads to a technique of "auditing," mentioned below.

On the other hand, confirmability is parallel to "objectivity" in conventional criteria, and is "concerned with establishing the fact that the data and interpretations of an inquiry were not merely figments of the inquirer's imagination." Researchers need to link "assertions, findings, and interpretations, and so on to the data themselves in readily discernible ways".

Although dependability and confirmability correspond to different notions in conventional criteria, both are realised by similar techniques in qualitative research. Of these techniques, "auditing" is emphasised strongly by many authors. Auditing in qualitative research is analogous to a fiscal audit. Schwandt (1997) states that auditing is "a procedure whereby a third-party examiner systematically reviews the audit trail maintained by the inquirer". In the case of qualitative interview research, the audit trail includes recorded materials such as cassette tapes, interview transcripts, interview guides, lists of interviewees, lists of categories and hypotheses the researcher used while analysing the data, notes about research procedures, and so on.

Lincoln and Guba (1985) claim that "even for a complex project, a week to ten days will be sufficient" to complete auditing. However, that may sounds too expensive for our readers, most of whom may be novices, and they might be relieved to see what other researchers have written:

The researcher's work in preparing an "audit trail" and the auditor's analysis, with its very detailed procedures, are at least as expensive.... We should probably expect that detailed documentation and auditing will continue to be restricted to high-stakes studies, or to those in which the researcher has a special interest in documentation or auditing as such.

Huberman and Miles (1994) also warn that this sort of re-analysis through auditing "raises questions about invasion of privacy and about potential harm to informants". In spite of these risks and limitations, simpler types of auditing could be useful tools for improving the quality of qualitative research.

2

Organizational Behaviour and Change

Organisation/Management theories developed from classical era to modern era. Development of various schools of thought in various era, throw light on development of behavioural issues in organisations.

Thorough understanding of this chapter would make your task easier, i.e., you can understand subsequent chapters in a synchronized way. To understand Organisational Behaviour, let us first take the two terms involved: organisation and behaviour. Organisation is a place where two or more people work together in a structured way to achieve a specific goal or set of goals. Goals are fundamental elements of organisations. According to Gary Johns, organisations are social interventions for accomplishing goals through group efforts. Various environmental forces influence organisations.

There are two types of environmental forces: direct and indirect. Some of the main direct forces are: customers, suppliers, competitors, labour market, and regulatory agencies. Some of the main indirect forces are: economic, technological, socio cultural, political, and international. Behaviour is anything that the human does.

Behaviour is response to stimulation that can be observed, thus it is any response or reaction of an individual. The basic unit of behaviour is activity. According to Luthans, in understanding the variable it is extremely important to separate the actual behaviour events from the outcomes of the events. Specific observable behavioural events and their patterns provide useful data in order to analyze the interaction, which precedes the behaviour and the consequences that follow the behaviour.

Running a large company, or even a small one, is no easy task. The field of Organisational Behaviour provides many helpful insights into understanding the complexities of people's behaviour on the job. Organisational Behaviour is the study and application of knowledge about how people act within organisations. The key elements in an organisation are: people, structure, technology, and external environment in which the organisation operates. When people join together in an organisation to accomplish an objective, some kind of structure is required. People also use technology to get the job done. So there is an interaction of people, structure, and technology. In addition, these elements are influenced by the external environment, and they influence it.

According to Keith Davis, Organisational Behaviour is an academic discipline concerned with understanding and describing human behaviour in an organisational environment. It seeks to shed light on the whole complex human factor in organisations by identifying causes and effects of that behaviour. According to Joe Kelly, Organisational Behaviour is the systematic study of the nature of organisations: how they begin, grow, and develop, and their effect on individual members, constituent groups, other organisations, and large institutions. According to Luthans, Organisational Behaviour is directly concerned with the understanding, prediction, and control of human behaviour in organisations.

According to Robbins, Organisational Behaviour is a field of study that investigates the impact that individuals, groups, and structure have on behaviour within organisations for the purpose of applying such knowledge towards improving an organisation's effectiveness. According to Baron and Greenberg, Organisational Behaviour is the field that seeks knowledge of behaviour in organisational settings by systematically studying individual, group, and organisational processes. On the basis of definitions stated and various other definitions, we can draw following conclusions related to nature and scope of Organisational Behaviour:

- Interdisciplinary Approach: Organisational Behaviour integrates knowledge from various relevant disciplines. This issue will be clear to you after reading the section on genesis of Organisational Behaviour in this unit.
- An Applied Science: Organisational Behaviour is oriented towards understanding the forces that affect behaviour so that their affects may be predicted and guided towards effective functioning of organisation. This issue will be clearer to you after reading the section on goals of Organisational Behaviour in this section.
- Behavioural Approach to Management: Organisational Behaviour is directly connected with the human side of management, but it is not the whole of management. Organisational Behaviour is related with the conceptual and human dimensions of management.
- Concern with Environment: Organisational Behaviour is concerned with issues like compatibility with environment e.g. person-culture fit, cross-cultural management etc.

- Scientific Method: Organisational Behaviour follows the scientific method and makes use of logical theory in its investigation and in answering the research questions. It is empirical, interpretive, critical and creative science.
- Contingency Approach: There are very few absolutes in Organisational Behaviour. The approach is directed towards developing managerial actions that are most appropriate for a specific situation.
- A Systems Approach: Organisational Behaviour is a systematic vision as it takes into account all the variables affecting organisational functioning.
- Value Centred : Organisational Behaviour is a value-centred science.
- Utilizes two Kinds of Logic: It utilizes both objective and subjective logic. Objectivity is concerned with reaching a fact through empirical analyses. Subjectivity is concerned with deciding about an issue through intuition, common sense, experiences, gut feeling, metaphors, learning from stories and cases, persuasive literature etc.

Organisational Behaviour focuses on five levels of analysis. They are :

- Individual behaviour
- Inter-personal behaviour
- Group behaviour and group dynamics
- Organisational issues
- Environmental issues.

STUDY OF ORGANIZATIONAL BEHAVIOUR

Organizational behaviour is a misnomer. It is not the study of how organizations behave, but rather the study of individual behaviour in an organizational setting. This includes the study of how individuals behave alone, as well as how individuals behave in groups.

The purpose of organizational behaviour is to gain a greater understanding of those factors that influence individual and group dynamics in an organizational setting so that individuals and the groups and organizations to which they belong may become more efficient and effective. The field also includes the analysis of organizational factors that may have an influence upon individual and group behaviour. Much of organizational behaviour research is ultimately aimed at providing human resource management professionals with the information and tools they need to select, train, and retain employees in a fashion that yields maximum benefit for the individual employee as well as for the organization.

Organizational behaviour is a relatively new, interdisciplinary field of study. Although it draws most heavily from the psychological and sociological sciences, it also looks to other scientific fields of study for insights. One of the main reasons for this interdisciplinary approach is because the field of organizational behaviour involves multiple levels of analysis, which are necessary to understand behaviour within organizations because people do not act in isolation. That is, workers influence their environment and are also influenced by their environment.

Individual Level of Analysis

At the individual level of analysis, organizational behaviour involves the study of learning, perception, creativity, motivation, personality, turnover, task performance, cooperative behaviour, deviant behaviour, ethics, and cognition. At this level of analysis, organizational behaviour draws heavily upon psychology, engineering, and medicine.

Group Level of Analysis

At the group level of analysis, organizational behaviour involves the study of group dynamics, intra- and intergroup conflict and cohesion, leadership, power, norms, interpersonal communication, networks, and roles. At this level of analysis, organizational behaviour draws upon the sociological and socio-psychological sciences.

Organization Level of Analysis

At the organization level of analysis, organizational behaviour involves the study of topics such as organizational culture, organizational structure, cultural diversity, inter-organizational cooperation and conflict, change, technology, and external environmental forces. At this level of analysis, organizational behaviour draws upon anthropology and political science. Other fields of study that are of interest to organizational behaviour are ergonomics, statistics, and psychometrics.

A number of important trends in the study of organizational behaviour are the focus of research efforts. First, a variety of research studies have examined topics at the group level of analysis rather than exclusively at the individual level of analysis. For example, while empowerment has largely been investigated as an individual-level motivation construct, researchers have begun to study team empowerment as a means of understanding differences in group performance. Similar research has focused on elevating the level of analysis for personality characteristics and cooperative behaviour from the individual level to the group level.

Another research trend is an increasing focus on personality as a factor in individual- and group-level performance. This stems from the movement toward more organic organization designs, increased supervisory span of control, and more autonomous work designs. All of these factors serve to increase the role that personality plays as a determinant of outcomes such as stress, cooperative or deviant behaviour, and performance. Personality traits that are related to flexibility, stress hardiness, and personal initiative are also the subject of research. Examples of these personality traits include a tendency toward individualism or collectivism, self-monitoring, openness to experience, and a proactive personality. Forms of behaviour that are constructive and change-oriented in nature are also studied. These forms of behaviour are proactive in nature and act to improve situations for the individual, group, or organization. Examples of these behaviors include issue selling, taking initiative, constructive change-oriented communication, innovation, and proactive socialization.

Other topics of interest in the field of organizational behaviour include the extent to which theories of behaviour are culturally bound, unethical decision-making, self-management and self-leadership, and work/family conflict.

ORGANIZATIONAL CHART

An organizational chart is a pictorial representation of a company's structure and reporting relationships. This chart can provide a great deal of information and may help organizational members understand the overall structure of the organization and its strategy. This entry describes how organizational charts are constructed, including the software that can be used to create them; what information the organizational chart provides; the benefits of making the chart available inside and outside of the organization; and the circumstances under which a chart is likely to change.

Constructing an Organizational Chart

All organizational charts have similar elements that allow them to be easily interpreted and understood by people inside and outside of the organization. Charts consist of shapes and lines that represent work units and their hierarchy.

The basic building block of an organizational chart is the rectangle, which can represent a person or a work unit (e.g., a department). If the outline of the rectangle is dashed, this means that a position is open and must be filled, as with one of the manager positions. If a rectangle is divided, and two or more names are in it, this may indicate job sharing or that multiple people are responsible for the outcomes associated with this position. W. Allen and P. Lloyd are comanagers in one area of the Production and Services Marketing Department, where they have a job sharing arrangement and each works part-time hours.

The boxes may contain as much or as little information as the organization prefers. They may include a job title, an employee's name, an employee's department, or even information such as job tenure, education, or salary. Alternatively, a chart may be created without rectangles, with names or titles standing alone. The three employees in the Public and Community Relations Department are listed with their names not in rectangles. This often is done to save space on the chart. Rectangles on an organizational chart are linked with solid or dashed lines. A solid line indicates a formal, direct relationship and a dashed line indicates that one employee or department advises another or has some other sort of indirect relationship. L. Jiminez has a dashed line to the Product and Services Marketing Department, which means that she sometimes will work for that department or will report to that department's manager. When lines represent a tree structure—when two or more rectangles are linked to another with multiple lines—this indicates that several individuals or departments report to one supervisor. For instance, the tree structure represents the relationship between the CEO and the three top managers who report to the CEO. Finally, a rectangle that is attached horizontally outside of the vertical hierarchy typically indicates an assistant or staff person. In the example, this is represented by the executive secretary to the CEO.

While organizational charts can be created by hand, most are created using computer software. Although it may be labour intensive, organizational charts can be created using drawing tools in a word processing program. Microsoft's Power Point presentation software allows for the creation of organizational charts, although there is little space available to create large charts. Specific software exists for creating larger, more complex charts, and there are many different packages available for purchase.

Some examples are OrgPlus5, ConceptDraw V, SmartDraw, and Abra OrgChart. These software programs allow for quick and easy chart creation with point-and-click menus and automatic resizing and alignment. Many of these software programs also allow one to easily download charts into a word processing document, a presentation, or a Web site. Other features available in these programs include the ability to insert employee photographs, as well as information from other human resources computer programs, directly into charts.

Information in the Organizational Chart

The organizational chart provides a great deal of information about the organization as a whole and the interaction of its parts. From a chart, one can see the organization's structure, its hierarchy, the degree to which it is centralized or decentralized, and its chain of command. Each of these is summarized below.

CLASSICAL ORGANIZATION THEORY

Classical organization theory evolved during the first half of this century. It represents the merger of scientific management, bureaucratic theory, and administrative theory. Frederick Taylor (1917) developed scientific management theory (often called "Taylorism") at the beginning of this century. His theory had four basic principles: 1) find the one "best way" to perform each task, 2) carefully match each worker to each task, 3) closely supervise workers, and use reward and punishment as motivators, and 4) the task of management is planning and control.

Initially, Taylor was very successful at improving production. His methods involved getting the best equipment and people, and then carefully scrutinizing each component of the production process. By analyzing each task individually, Taylor was able to find the right combinations of factors that yielded large increases in production.

While Taylor's *scientific management theory* proved successful in the simple industrialized companies at the turn of the century, it has not faired well in modern companies. The philosophy of "production first, people second" has left a legacy of declining production and quality, dissatisfaction with work, loss of pride in workmanship, and a near complete loss of organizational pride.

Max Weber (1947) expanded on Taylor's theories, and stressed the need to reduce diversity and ambiguity in organizations. The focus was on establishing clear lines of authority and control. Weber's *bureaucratic theory* emphasized the need for a hierarchical structure of power. It recognized the importance of division of labor and specialization. A formal set of rules was bound into the hierarchy structure to insure stability and uniformity. Weber also put forth the notion that organizational behavior is a network of human interactions, where all behavior could be understood by looking at cause and effect.

Administrative theory (i.e., principles of management) was formalized in the 1930's by Mooney and Reiley (1931). The emphasis was on establishing a universal set of management principles that could be applied to all organizations. Classical management theory was rigid and mechanistic. The shortcomings of classical organization theory quickly became apparent. Its major deficiency was that it attempted to explain peoples' motivation to work strictly as a function of economic reward.

Neoclassical Organization Theory

The human relations movement evolved as a reaction to the tough, authoritarian structure of classical theory. It addressed many of the problems inherent in classical theory. The most serious objections to classical theory are that it created overconformity and rigidity, thus squelching creativity, individual growth, and motivation. Neoclassical theory displayed genuine concern for human needs. One of the first experiments that challenged the classical view was conducted by Mayo and Roethlisberger in the late 1920's at the Western Electric plant in Hawthorne, Illinois (Mayo, 1933). While manipulating conditions in the work environment (e.g., intensity of lighting), they found that any change had a positive impact on productivity. The act of paying attention to employees in a friendly and nonthreatening way was sufficient by itself to increase output. Uris (1986) referred to this as the "wart" theory of productivity. Nearly any treatment can make a wart go away—nearly anything will improve productivity. "The implication is plain: intelligent action often delivers results" (Uris, 1986, p. 225).

The Hawthorne experiment is quite disturbing because it cast doubts on our ability to evaluate the efficacy of new management theories. An organization might continually involve itself in the latest management fads to produce a continuous string of Hawthorne effects. "The result is usually a lot of wheel spinning and cynicism" (Pascale, 1990, p. 103). Pascale believes that the Hawthorne effect is often misinterpreted. It is a "parable about researchers (and managers) manipulating and 'playing tricks' on employees." (p. 103) Erroneous conclusions are drawn because it represents a controlling and manipulative attitude toward workers.

Writing in 1939, Barnard (1968) proposed one of the first modern theories of organization by defining organization as a system of consciously coordinated activities. He stressed in role of the executive in creating an atmosphere where there is coherence of values and purpose. Organizational success was linked to the ability of a leader to create a cohesive environment. He proposed that a manager's authority is derived from subordinates' acceptance, instead of the hierarchical power structure of the organization. Barnard's theory contains elements of both classical and neoclassical approaches. Since there is no consensus among scholars, it might be most appropriate to think of Barnard as a transition theorist.

Simon (1945) made an important contribution to the study of organizations when he proposed a model of "limited rationality" to explain the Hawthorne experiments. The theory stated that workers could respond unpredictably to managerial attention. The most important aspect of Simon's work was the rigorous application of the scientific method. Reductionism, quantification, and deductive logic were legitimized as the methods of studying organizations.

Taylor, Weber, Barnard, Mayo, Roethlisberger, and Simon shared the belief that the goal of management was to maintain equilibrium. The emphasis was on being able to control and manipulate workers and their environment.

Contingency Theory

Classical and neoclassical theorists viewed conflict as something to be avoided because it interfered with equilibrium. Contingency theorists view conflict as inescapable, but manageable.

Chandler (1962) studied four large United States corporations and proposed that an organization would naturally evolve to meet the needs of its strategy — that form follows function. Implicit in Chandler's ideas was that organizations would act in a rational, sequential, and linear manner to adapt to changes in the environment. Effectiveness was a function of management's ability to adapt to environmental changes.

Lawrence and Lorsch (1969) also studied how organizations adjusted to fit their environment. In highly volatile industries, they noted the importance of giving managers at all levels the authority to make decisions over their domain. Managers would be free to make decisions contingent on the current situation.

Systems Theory

Systems theory was originally proposed by Hungarian biologist Ludwig von Bertalanffy in 1928, although it has not been applied to organizations until recently (Kast and Rosenzweig, 1972; Scott, 1981).

The foundation of systems theory is that all the components of an organization are interrelated, and that changing one variable might impact many others. Organizations are viewed as open systems, continually interacting with their environment. They are in a state of dynamic equilibrium as they adapt to environmental changes.

Senge (1990) describes systems thinking as:

understanding how our actions shape our reality. If I believe that my current state was created by somebody else, or by forces outside my control, why should I hold a vision? The central premise behind holding a vision is that somehow I can shape my future, Systems thinking helps us see how our own actions have shaped our current reality, thereby giving us confidence that we can create a different reality in the future. (p. 136)

A central theme of systems theory is that nonlinear relationships might exist between variables. Small changes in one variable can cause huge changes in another, and large changes in a variable might have only a nominal effect on another. The concept of nonlinearity adds enormous complexity to our understanding of organizations. In fact, one of the most salient argument against systems theory is that the complexity introduced by nonlinearity makes it difficult or impossible to fully understand the relationships between variables.

Organizational Structure

Until recently, nearly all organizations followed Weber's concept of bureaucratic structures. The increased complexity of multinational organizations created the necessity of a new structure that Drucker called (1974) "federal decentralization". In federal decentralization, a company is organized so that there are a number of independent units operating simultaneously. "Each unit has its own management which, in effect, runs its own autonomous business." (p. 572) This structure has resulted in large conglomerates which have diversified into many different fields in order to minimize risk.

The project management organizational structure has been used effectively in highly dynamic and technological environments (French, Kast and Rosenzweig, 1985). The project manager becomes the focal point for information and activities related to a specific project. The goal is to provide effective integration of an organization's resources towards the completion of a specific project. Impementing a project management approach often involves dramatic changes in the relationships of authority and responsibility. The matrix organizational structure evolved from the project management form (Kolodny, 1979). It represents a compromise between the traditional bureuacratic approach and the autonomous project management approach. A matrix organization has permanently established departments that provide integration for project management. The matrix form is superimposed on the hierarchical structure, resulting in dual authority and responsibilities. Permanent functionality departments allocate resources to be shared among departments and managers.

Systems theory views organizational structure as the "established pattern of relationships among the parts of the organization" (French, Kast, and Rosenzweig, 1985, p. 348). Of particular importance are the patterns in relationships and duties. These include themes of 1) integration (the way activities are coordinated), 2) differentiation (the way tasks are divided), 3) the structure of the hierarchical relationships (authority systems), and 4) the formalized policies, procedures, and controls that guide the organization (administrative systems).

The relationship between the environment and organizational structure is especially important. Organizations are open systems and depend on their environment for support. Generally, more complex environments lead to greater differentiation. The trend in organizations is currently away from stable (mechanistic) structures to more adaptive (organic) structures. The advantage is that organizations become more dynamic and flexible. The disadvantage is that integration and coordination of activities require more time and effort.

The relationship between an organization and its environment is characterized by a two-way flow of information and energy. Most organizations attempt to influence their environment. Advertising campaigns and lobbying efforts are two examples. Some theorists believe that "... environments are largely invented by organizations themselves. Organizations select their environments from ranges of alternatives, then they subjectively perceive the environments they inhabit" (Starbuck, 1976, p. 1069). Strategic decisions regarding product lines and distribution channels contribute to the selection of the organizational structure and the environment. It is a commonly held tenant that people are less satisfied with their work in highly structured organizations. Many research studies have been conducted to examine the relationship between organizational structure and employee behavior (e.g., satisfaction, performance, and turnover). However, the results of these studies are contradictory (Dalton, et al., 1980). Structural deficiencies can result in low motivation and morale, decisions lacking in timeliness or quality, lack of coordination and conflict, inefficient use of resources, and an inability to respond effectively to changes in the environment (French, Kast, and Rosenzweig, 1885).

One enduring and controversial debate about organizational structure is whether or not there is a maximum desirable size for an organization, after which there will be declining effectiveness. Does an organization become increasingly dysfunctional as it exceeds its "ideal" size? Several researchers have hypothesized that organizational growth is beneficial only up to a point. Most researchers support a curvilinear growth theory. Pfeffer and Salancik (1978) found that profitability increases with size and then tapers off. Warwick (1975) reported that the growth in the U.S. State Department resulted in decreased flexibility and responsiveness, even though specific steps had been taken to abate these problems. There are several theories to explain these findings. The most common explanation is based on the fact that an organization's size is usually positively correlated with age. Older (i.e., larger) organizations have become more rigid in their ways and they are less able to adapt to change. Another popular theory is that in larger organizations, workers' jobs become more specialized. The lack of variety creates a less motivating environment. Other theories have proposed that excessive size creates crippling coordination problems.

ORGANIZATIONAL BIRTH AND GROWTH

Clearly, one of the most dominant themes in the literature has been to define organizations from the perspective of their position on a growth curve. Cameron and Whetten (1983) reviewed thirty life-cycle models from the organizational development literature. They summarized the studies into an aggregate model containing four stages. The first stage is "entrepreneurial", characterized by early innovation, niche formation and high creativity. This is followed by a stage of "collectivity", where there is high cohesion and commitment among the members. The next stage is one of "formalization and control", where the goals are stability and institutionalization.

The last stage is one of "elaboration", characterized by domain expansion and decentralization. The striking feature of these life-cycle models is that they did not include any notion of organizational decline. They covered birth, growth, and maturity, but none included decline or death. The classic S-curve typifies these life-cycle models. Whetten (1987) points out that these theories are a reflection of the 1960s and 1970s, two highly growth oriented decades.

Land and Jarman (1992) have attempted to redefine the traditional S-curve that defines birth, growth, and maturity. The first phase in organizational growth is the entrepreneurial stage. The entrepreneur is convinced that their idea for a product or service is needed and wanted in the marketplace.

The common characteristic of all entrepreneurs and new businesses is the desire to find a pattern of operation that will survive in the marketplace. Nearly all new businesses fail within the first five years. Land and Jarman (1992) argue that this is "natural", and that even in nature, cell mutations do not usually survive. This phase is the beginning of the S-curve.

The second phase in organizational growth is characterized by a complete reversal in strategy. Where the entrepreneurial stage involves a series of trial and error endeavors, the next stage is the standardization of rules that define how the organizational system operates and interacts with the environment. The chaotic methods of the entrepreneur are replaced with structured patterns of operation. Internal processes are regulated and uniformity is sought. During this phase, growth actually occurs by limiting diversity. "Management procedures, processes, and controls are geared to maintain order and predictability" (Land and Jarman, 1992). This phase is the rapid rise on the S-curve.

Organizational growth does not continue indefinitely. An upper asymptopic limit can be imposed by a number of factors. Land and Jarman (1992, p. 258) identify the most common reasons why organizations reach upper growth limits:

- Rapidly increasing internal and market place complexity in such areas a product proliferation and market divisions
- Internal competition for resources
- Increasing cost of manufacturing and sales
- Diminishing returns

- Declining share of the market
- Decreasing productivity gains
- Growing external pressures from regulators and influence groups
- Increasing impact of new technologies
- New and unexpected competitors.

The transition to the third phase involves another radical change in an organization. Most organizations are not able to make these changes, and they do not survive. "The organization must open up to permit what was never allowed in to become a part of the system, not only by doing things differently, but by *doing different things*" (Land and Jarman, 1992, p. 257). The organization needs to continue its core business, while at the same time engaging in inventing new business. This bifurcation is necessary because the entrepreneurial environment (of inventing business) is incompatible with the controlling environment of the core business.

The goal is a continuing integration of the new inventions into the mainstream business, where a re-created organization emerges. The core business is changed by the inventions it assimilates, and the organization takes on a new form. Land and Jarman (1992) believe that the greatest challenge facing today's organizations is the transition from phase two to phase three. "Organizations defeat their best intentions by continuing to operate with essential beliefs that automatically perpetuate the second phase." (p. 264)

There are several factors that contribute to organizational growth (Child and Kieser, 1981). The most obvious is that growth is a by-product of another successful strategy. A second factor is that growth is deliberately sought because it facilitates management goals. For example, it provides increased potential for promotion, greater challenge, prestige, and earning potential. A third factor is that growth makes an organization less vulnerable to environmental consequences. Larger organizations tend to be more stable and less likely to go out of business (Caves, 1970; Marris and Wood, 1971; Singh, 1971). Increased resources make diversification feasible, thereby adding to the security of the organization.

Child and Kieser (1981) suggest four distinct operational models for organizational growth. 1) Growth can occur within an organization's existing domain. This is often manifest as a striving for dominance within its field. 2) Growth can occur through diversification into new domains. Diversification is a common strategy for lowering overall risk, and new domains often provide fertile new markets. 3) Technological advancements can stimulate growth by providing more effective methods of production. 4) Improved managerial techniques can facilitate an atmosphere that promotes growth. However, as Whetten (1987) points out, it is difficult to establish cause and effect in these models. Do technological advancements stimulate growth, or does growth stimulate the development of technological breakthroughs? With the lack of controlled experiments, it is difficult to choose between the chicken and the egg.

Organizational Decline

Until recently, most theories about organization development viewed decline as a symptom of ineffective performance. Well-managed organizations were expected to grow year after year. Implicit in these theories was the idea that organizational growth is synonymous with expansion. These theories reflected what scholars observed in the business world. Organizational growth was an indicator of successful management.

Kenneth Boulding (1950) proposed a biological model of economics, characterized by birth, maturation, decline, and death. He argued that in all organisms, there is an "inexorable and irreversible movement towards the equilibrium of death." (p. 38) Many organizational theorists took strong exception to Boulding's biological determinism theory. They maintained that organizations are not constrained by a defined life cycle, and there is no indication that all organizations need to die. The 1980's ushered in a new era where organizational decline was apparent everywhere. Management strategies involved reducing employees, salary freezes and reductions, cutting administrative overhead, and consolidating operations. It became clear that the traditional S-curve model was incomplete and did not address the issues of declining organizations.

One of the problems in the literature is that it is difficult to agree on a precise definition of organizational decline. Is a company in decline when it cuts back the number of employees in order to become more profitable? A common definition of decline is a decrease in profit or budget. Most theorists agree that decline negatively impacts individuals and the organization as a whole. Cameron, Whetten, and Kim (1987) argue that decline results in decreased morale, innovativeness, participation, leader influence, and long-term planning. They associate decline with, conflict, secrecy, rigidity, centralization, formalization, scapegoating, and conservatism.

Nystrom and Starbuck (1984) attribute organizational decline to over-confidence. According to this theory, a successful past can lure an organization to become over-confident in its ability to prosper. This leads to a lackadaisical attitude towards new innovations, quality, and customer satisfaction. Another theory is that large size promotes rigidity, which makes it cumbersome for an organization to respond to environmental changes (Whetten, 1987).

In applying the biological life-cycle model to organizations, Wilson (1980) identified two different types of organizational decline: "k" and """r" extinction.

When an organization has reached the upper asymptopic limit defined by carrying capacity of its niche, it declines because of k-extinction. The organization has exhausted its environmental resources, or other organizations have begun competing for limited resources. When an organization falls short of its upper asymptopic limit, and begins declining without reaching its maximum potential, it is called r-extinction. Bad management or a failure to remain competitive are the most common reasons for r-extinction.

Bibeault (1982) proposed a four-stage model to describe the process of turning around an organization in decline. The key to the process was to replace the top personnel. Bibeault argued that only way to reverse a decline is to 1) change the management, the rationale being that "problem causers have little credibility as problem solvers" (Whetten, 1987, p. 37). Chaffee (1984) also stressed the symbolic value of changing administrative personnel. Change in management is followed by 2) an evaluation stage, 3) implementing emergency actions and stabilization procedures, and finally, 4) a return to growth.

A different approach for describing organizational turnaround was proposed by Zammuto and Cameron (1985). Their model was based on the idea that turnaround could be accomplished by addressing five process domains. 1) The defense domain involves strategies for protecting the organization from a hostile environment. An example would be an organization that forms a common-purpose coalition with other organizations. 2) The offense domain involves expanding on the activities that the organization already does well. 3) Creating new domains consists of diversification activities. 4) The consolidation domain involves reducing the scope of activities by cutting back to core products and services. 5) The substitution domain involves replacing one set of activities with another.

In contrast to these theories, Harrigan (1980, 1981, 1982) and Porter (1980) have looked at how organizations respond to decline as a result of environmental limitations (i.e., k-extinction). Organizational activities often involve attempts to focus on a specific market niche in which the organization might have a competitive advantage. Another approach is to rapidly liquidate the organization, and extract as much remaining value as possible, although Harrigan (1982) notes that there are often financial, legal, structural, and emotional obstacles to this strategy.

The most common response to organizational decline is retrenchment. Whetten (1987) identifies three sequential stages involved in the process. The first is one of identification. Management must be sensitive to problems when they first appear, and be able to meet the problems head on. The second is one of communication. Management must communicate a clear message of the organization's situation and instill confidence in its ability to meet the crisis. The third stage involves the implementation of a downsizing program.

Sutton (1983) surveyed managers to examine their beliefs regarding how employees would react to an organizational closing. It was found that managers had several inaccurate perceptions. For example, managers' incorrectly believed that productivity and quality would plummet, employee sabotage and theft would increase, and there would be increases in conflict. On the other hand, Sutton's study did offer evidence that rumors were abundant, the best employees sought different employment, and that employee's had trouble accepting the closing.

THE LEARNING ORGANIZATION

Peter Senge (1990) defines learning as enhancing ones capacity to take action. "So learning organizations are organizations that are continually enhancing their capacity to create." (p. 127) Senge believes that organizations are evolving from controlling to predominantly learning.

Senge (1990) discusses learning disabilities in companies. One of the most serious disabilities is when people form a strong identification with their position. What they do becomes a function of their position. They see themselves in specific roles, and are unable to view their jobs as part of a larger system. This often leads to animosity towards others in the organization, especially when things go wrong. Another disability is that we are slow to recognize gradual changes and threats.

Senge (1990) refers to several other learning disabilities as "myths". He discusses the "myth of proactiveness", where "proactiveness is really reactiveness with the gauge turned up to 500%." (p. 129) Another myth is that we "learn from experience". Senge maintains that we actually only learn when the experience is followed by immediate feedback. Another myth is that management teams can provide creative and beneficial solutions. Senge maintains that the result of management teams is "skilled incompetence, where groups are highly skilled at protecting themselves from threat, and consequently keeping themselves from learning." (p. 131)

Senge (1990) believes that new organizations can be built by adopting a set of disciplines, where a discipline is defined as a "particular theory, translated into a set of practices, which one spends one's life mastering." (p. 131) Thus, mastering a discipline becomes a life-long learning process. According to Senge, there are five disciplines important to the learning organization. The first discipline is "building a shared vision". "Building" involves an ongoing process, and "shared" implies that the vision is held in common by individuals. A second discipline of "personal mastery" demonstrates a commitment to the vision. A third discipline involves the idea of mental models, where we construct internal representations of reality. An important element of using mental models is the need to balance inquiry and advocacy. A fourth discipline in that only shared mental models are important for organizational learning. The fifth discipline is a commitment to a systems approach.

Community

Gozdz (1992) believes that learning organizations are centered around the concept of community. "An organization acting as a community is a collective lifelong learner, responsive to change, receptive to challenge, and conscious of an increasingly complex array of alternatives." (p. 108) Communities provide safe havens for its members and foster an environment conducive to growth. Gozdz describes the community as group of people who have a strong commitment to "ever-deepening levels of communication." (p. 111)

M. Scott Peck (1987) describes the process of building a community in *The Different Drum*. An organization goes through a four-stage process. The first stage is one of denial. Group members ignore differences in power, and pretend that they are a community. Decision-making processes go unchallenged. The next stage occurs when differences between members become apparent. Attempts are made to restore the situation to what has worked in the past by eliminating differences. An organization enters the third stage when members realize that their efforts to control differences have failed. They begin communicating and true collaborative efforts emerge. In the final stage, there is the true spirit of community. Differences are embraced. Decisions are made collectively. Learning and innovation comes from the group as a whole

Many organization experience brief periods of community, but they are not able to sustain those periods. Gozdz (1992) describes this failure as a lack of discipline and commitment.

There is an illusion that once a sense of community occurs within an organization it will remain constant. This is not the case. The sense of community or flow state is repeatedly lost. It can be deliberately regained at ever greater levels of organizational maturity, but only when sustaining community is seen and accepted as a path to developing mastery. This path is community as a discipline. (p. 114)

According to Gozdz (1992), the job of the leaders in the process of community building is to keep peoples' attention focused on the process. The four stages of community development are repeated over and over again.

New situations and contingencies arise that initiate new cycles in the growth process.

ORGANIZATIONAL MORALITY

The classical view of organizational responsibility is best illustrated by Adam Smith's (1937) belief that an "invisible hand" directs all activities towards the public good, and that the responsibility of an organization was only to maximize profits within the constraints of the law. The free market system was seen as a self-controlling mechanism, whereby an organization producing the best goods and services would prosper. Any interference with the free market system was viewed as an affront against the best interests of society.

The accountability concept states that organizations receive their charter from society as a whole, and therefore their ultimate responsibility is to society. Environmental and worker protection laws reflect the belief that maximization of profits is secondary to the health of society. The extensive proliferation of laws restricting business demonstrates a growing skepticism concerning the morality and ethics of corporate management.

Some theorists believe that organizations have the social responsibility "to take actions which protect and improve the welfare of society as a whole along with their own interests" (Davis and Blomstrom, 1980, p. 6). Others take a more narrow approach, and believe that social responsibility extends only to "social problems caused wholly or in part by the corporation" (Fitch, 1971, p. 38).

Linda Stark (1989) discusses the five stages of corporate moral development, although she is quick to point out that progression through the stages is neither linear or one direction. An amoral corporation pursues profit at any cost. A legalistic corporation follows the letter of the law, but not the spirit. A responsive corporation makes ethical decisions based on long-term economic decisions. An emergent ethical corporation recognizes its social responsibility and balances ethics and profitability. The ethical corporation places social responsibility at its center and bases its existence on ethics.

Environmental awareness has evolved to become a major ethical consideration in many corporations. During the 1950's, science and technology were viewed as the answer to the world's problems. The ecological ramifications of that era became apparent in the 1960's. The 1970's began with the organization of the first Earthday. The Environmental Protection Agency (EPA) and the Occupational Safety and Health Administration (OSHA) were created to monitor the environment and worker safety. During the 1980's, many corporations began to take proactive conservation measures. Environmental considerations began to be addressed at the manufacturing level so that harmful materials and waste were minimized or removed from the production process. Citizen action groups became increasingly effective in forcing corporations to examine their environmental impact. In the 1990's, many corporations have adopted the policy of "sustainable development." The key issue is that environmental protection is one of the highest priorities of every business.

GOALS OF ORGANISATIONAL BEHAVIOUR

Field of OB faces a special challenge. In the areas of physical science, accounting, mathematics etc. if you do not know a concept, you would not claim that you know it. However in the field of human behaviour, though we may not know a fact, yet through our accumulated knowledge it may appear that we know it and, in this long drawn conclusion you may be far away from the fact. For example it appears that high job satisfaction would necessarily lead to high organisational commitment but most of the studies have stood against this apparently obvious hypothesis. One of the objectives of a course in OB is to replace popularly held notions, often accepted without question, with science-based conclusions.

Since 1950s till date hundreds of thousands of research studies have been done on various aspects of OB, and several hundreds of research studies still continue to investigate facts. OB attempts to test theories through scientific research process. Once a theory has been formulated, predictions derived from it are tested through direct research.

If these are confirmed, confidence in the theories is increased. If they are disconfirmed, confidence is diminished. At this point, the theory is either modified and retested, or completely rejected. Theory building and empirical research coexist and reinforce each other. A good theory has to be of practical use and empirical validation would confirm this. Likewise, a good empirical research should have its foundation in a viable theory and should add to the body of existing knowledge.

There are mainly three goals of OB:

- Understanding behaviour
- Which variables are important?
- How strong are they?
- How do they interrelate?
- Predicting behaviour
- What patterns of behaviour are present?
- What is the cause effect relationship?
- Controlling behaviour
- What solutions are possible?

- Which variable can be influenced?
- How can they be influenced?

Traditional and New Approaches to OB

Study of human behaviour, being a part of general management, can be traced back to 4,000 B.C., when the Egyptian pyramids were built or even the dawn of mankind when people hunted in groups and protected their families or communities against hostile environmental forces. However for the purpose of our study we need to evaluate how OB developed during last two centuries.

In Unit 2 you have already studied about various theories related to development of organisation as well as concept of management. We can say here that along with changes in theories of organisation, our understanding of OB also developed.

- The Scientific Management Theories and the School: mainly developed by F.W.Taylor, H.L.Gnatt, Frank and Lillian Gilberth.
- Classical Organisation Theory School: mainly developed by Henri Fayol, Max Weber, Mary Parker Follet, Chester Barnard.
- The Human Relations School: mainly developed by Elton Mayo through Hawthorne Experiments.
- The Behavioural Science School: mainly developed by Abraham Maslow, Douglas McGregor, James March, Herbert Simon.
- The Systems Approach.
- The Contingency Approach.
- Contemporary Approach.

After studying the historical development of organisational theories, you might have noticed that with passage of time following issues occurred:

- Human factor became more important successively.
- Focus shifted from individual performance to both individual as well as group (team) performance.
- Emphasis given on actualizing the human potential.
- Emphasis on developing managerial and human skills on continuous basis.
- Emphasis on human relationship and informal organisation.
- Emphasis on creating synergy through team work.
- Treating employees with more dignity as a wholesome person.
- Increasing importance to environmental factors influencing organisation.
- Importance to psychological contract between individuals and organisation.
- Increasing concern for people in organisation.

- Continuous effort to establish effective organisational culture and climate.

According to Robert Baron, four major features characterise modern OB:

- It has adopted a somewhat more positive view of human being in work settings than prevailed in the past.
- By drawing on several related fields, it has attained a high degree of sophistication about human behaviour.
- It has adopted a contingency approach to behaviour in organisation-assuming that there is nothing like permanent way of arriving at a particular solution effectively.
- It is integrative in nature. It seeks to comprehend bahaviour in organisations by combining information from several different levels of analysis.

Apart from the description above, modern OB is concerned with the issues like: managing intelligence quotient, emotional quotient, and spiritual quotient, improving mental hygiene and overall health of members, continuous improvement of skills and values through training, managing ethical practices, accomplishing fulfilment of psychological contract between individuals and the organisation, quest for quality, behavioural intervention in merger and acquisition as well as in rightsizing, cross-culture management, managing multinational organisations etc.

According to modern thoughts on OB, it is necessary to understand the interrelationships between human behaviour and other variables, which together comprise the total organisation. These variables provide parameters within which a number of interrelated dimensions can be identified-the individual, the group, the organisation, and the environment-which collectively influence behaviour in work organisations.

MANAGING CHANGE IN ORGANIZATION

Organization theory is a broad field with roots in sociology. Anthropologists, philosophists, and political scientists have contributed greatly to the field. Organization theory as a topic for managers, as opposed to scholars, has come about fairly recently. For example, political scientists trace many ideas back to Ancient Rome or even before. Philosophers reach even farther back in time, and anthropologists have been interested in organization in terms of how groups arrange social systems and status systems as long as there has been a field of anthropology.

Although it is difficult to pinpoint when managers became interested in theories of organization, some suggest it was around the 1940s, when the writings of the German sociologist and engineer, Max Weber, were translated into English. Although Weber's life-work was spent trying to understand why capitalism arose first in Western Europe rather than in Asia, American managers lifted Weber's notion of bureaucracy out of these studies to explain then-modern forms of organizations that coincided with what Weber had described among Western European organizations. These organizations had reduced the influence of patriarchal styles of management to systems in which job positions, rather than the people in those positions, provided the source of authority.

Phillip Selznick became well known for studying goal conflict and power struggles during the U.S. government's subsidy of the Tennessee Valley Authority, which was an effort to put unemployed Americans to work producing electricity for the Tennessee Valley area. In the 1950s Herbert Simon studied how managers make decisions when information is complex but incomplete. The 1960s brought about research elaborating why the closed-system mentality of organizations—the idea that organizations have little reciprocal interaction with their environments—was not accurate. The result was a shift to viewing organizations as open systems that are intertwined with their environments in such a way that reciprocal interdependence is created. Managers began to realize that society has a profound effect on organizations, as organizations also have a profound effect on society.

Taken in total, then, organization theory as a management topic of interest was born out of non-management research. Bureaucracy, authority, goal conflict, power, managerial decision-making, and interaction between organizations and their environments are all topics of concern among today's organization theorists, but none of these ideas arose from management research.

Management research has borrowed from these various fields to attempt to answer two questions that are of specific interest to managers: Why do organizations exist and how do they function? With respect to the first question, it seems reasonable to argue that organizations exist to provide society with a level of goods and services that would otherwise be unattainable. With respect to the second, how organizations function, organizations combine human skills, knowledge, technology, and material resources to produce goods and services. The broadest variant of organization theory looks at the relationship between organizations and their social and natural environments, as originated by open-system theorists. One branch of Western moral philosophy conceives of organizations as having a social contract with society, whereby they are granted legitimacy for the purpose of serving the social good. This constructive purpose includes the production of goods and services and their fair allocation. Yet organizations can also cause negative social outcomes. Negative externalities include the social problems associated with monopolies, unsafe products, and the unfair treatment of employees, as well as the ecological problems posed by industrial accidents and industry-related pollution of the natural environment.

Because of the existence of negative externalities, organization theory is influenced by the concept of social control that can be found in political science, economics, and sociology. Social control refers to the laws, regulations, and social customs that are meant to minimize the negative impacts of organizational activities. Because organization theorists address social control, they also examine the nature of the relationship between organizations and their regulatory agencies.

The anthropological view derives from the knowledge that standards of social control reflect the underlying assumptions, values, and beliefs of cultures. The restraints or expectations placed on organizations, therefore, can vary across societies. The stakeholder model of organizations is useful for demonstrating this form of cultural relativism. The stakeholder model depicts an organization as surrounded by a variety of constituent groups, such as customers, social activists, regulators, the media, stockholders, and regulatory agencies.

Stakeholder expectations, in turn, can depend upon cultural affiliations. For example, employees in the United States tend to expect more active participation in the work contract than do employees in Japan. Thus, a Japanese firm with operations in the United States may face employment laws and employee views of justice and fairness that differ from those of the home country. When the scope of organization theory is widened to include international issues such as the activities of multinational corporations in host environments, the impact of cultural relativism must be acknowledged.

Although organization theory never loses sight of the importance of the organizational-societal interface (and this is one distinguishing feature between organization theory and organization behaviour), it also deals with what goes on inside organizations. Weber's classical theory of bureaucracy, for example, was followed by research on the more informal aspects of organizational life. This line of inquiry is strongly influenced by the insight that organizational activity can involve less-than-rational processes that yield unexpected consequences, including the negative externalities mentioned above. Hence, although organizations ideally exist as tools for constructive social purposes, these purposes can be subverted by the constraints on rational decision processes within organizations.

Contemporary theorists use many metaphors to guide their investigations into the suboptimal and even non-rational decisions enacted in organizations. These metaphors include those that suggest organizations are systems of power and political intrigue, miniature cultures, chaos, temples, theatres, machines, families, and jungles. These metaphors arise through shared meaning that has been socially constructed and generally agreed upon, thus subconsciously institutionalizing the specific, agreed-upon metaphor for certain organizations. Organization theorists are interested in why organizations exist and how these social systems function. This interest has yielded a body of research on the organizational-societal relationship and the formal and informal aspects of organizational life, yet there is no single answer to either of these root issues.

ORGANIC ORGANIZATIONS

The term "organic" suggests that, like living things, organizations change their structures, roles, and processes to respond and adapt to their environments. Burns and Stalker noted in *The Management of Innovation* that organic structures are appropriate in unstable, turbulent, unpredictable environments and for non-routine tasks and technologies. For organizations coping with such uncertainty, finding appropriate, effective, and timely responses to environmental challenges is of critical importance. Organic organizations are characterized by:

- decentralization
- flexible, broadly defined jobs
- interdependence among employees and units
- multi-directional communication
- employee initiative
- relatively few and broadly defined rules, regulations, procedures, and processes
- employee participation in problem solving and decision making, often interactively and in groups

In organic organizations, the emphasis is on effectiveness, problem solving, responsiveness, flexibility, adaptability, creativity, and innovation. Such an organization is able to respond in a timely manner to environmental change because employees are empowered to be creative, to experiment, and to suggest new ideas. The process of innovation is triggered by employees throughout the organization in a "bottom-up" manner. These characteristics fit together in a cohesive organizational structure that allows for flexibility and ongoing change.

Meeting Challenges

An unstable external environment increases the uncertainty and complexity with which an organization must contend. An organization is continually confronted with a variety of new and unexpected problems and opportunities, of which the nature and relevant factors are initially unclear and for which appropriate responses are not immediately obvious.

Further, since the environment changes rapidly, responses to today's problems and opportunities may need to be modified or may even be inappropriate or irrelevant to tomorrow's challenges. In short, the organization cannot keep doing the same old things in the same old ways. Under conditions of uncertainty and complexity, the organization must design its structures and processes to be flexible and responsive to changes in customer desires, technology, governmental regulations, and economic conditions.

Flexibility and Shared Authority

The need for flexibility and responsiveness leads to the decentralization of decision-making authority in organic organizations. As a result, rules, regulations, procedures, and policies tend to be few, are defined broadly rather than precisely, loosely rather than rigidly, and are often informal rather than written. Employees are allowed to exercise a great deal of discretion.

The authority to identify problems and opportunities and to devise responses is delegated to those best able to respond, regardless of their position, unit, or level in the organization. Emphasis is placed more on individual and group control than on managerial, hierarchical control. Top-level managers in organic organizations are more concerned with coordination and integration as opposed to passing directives down a vertical hierarchy, which is a common task of top-level managers in mechanistic organizations.

The need for flexibility and responsiveness also affects how work is designed and performed in organic organizations. Jobs are not clearly or precisely defined in these organizations. Positions, roles, job descriptions, and standard operating procedures are broad and generalized rather than specific and specialized. Employees accept general responsibility for getting things done, but the manner in which they accomplish their tasks is dictated more by autonomous or semi-autonomous teams than by standard operating procedures. Because the work of organic organizations is often interdependent, specific tasks and responsibilities vary from one situation to another and are refined through direct interaction and mutual adjustment among employees and work units. Too much direction from top-level management may hinder rather than assist the accomplishment of tasks.

A key issue in organic organizations is determining who has the knowledge, perspective, experience, expertise, or skills required to identify opportunities or find solutions to problems. Rather than assuming that top management is the fountainhead of all knowledge and wisdom, organic organizations assume that various people in the organization may have crucial insights or capabilities.

Thus, communication is multidirectional, decentralized, and informal rather than hierarchical and formalized. In order to facilitate the sharing of information and ideas, employees are frequently empowered to communicate across traditional organizational boundaries regardless of position or level or unit.

Going one step further, pharmaceutical firms, for example, may collaborate across corporations and with academic researchers to conduct basic research leading to new drug development. Jack Welch, former CEO of General Electric, referred to this type of company as a "boundaryless organization." Coordination and integration with multiple constituencies beyond traditional organizational boundaries is a necessary component for success, especially in multinational organizations.

Diversity of information and perspectives is often the key to the development of creative responses to vague, complex problems and opportunities. Thus, in organic organizations, much work is done in groups composed of employees with different backgrounds and from different levels, units, or functional areas. Such teams are among the main coordination mechanisms in organic organizations.

The Human Element

Human needs and dynamics play an important role in organic organizations. The empowerment and participation of employees is motivational because it meets the human need for autonomy, responsibility, challenge, esteem, social interaction, and personal development. Furthermore, this empowerment and participation helps the organization develop and capitalize on its intellectual capital, which is becoming increasingly valued by many organizations. By emphasizing initiative, direct interaction, open communication, and the creation of teams composed of various members of the organization, organic organizations are able to utilize their internal diversity to foster innovative responses to environmental challenges and changes.

Mixing Styles

The organic organization is not entirely without hierarchy or formalized rules, regulations, procedures, and processes. Indeed, structural parameters, even if loosely or broadly defined, are necessary to prevent the chaos that would result from absolute decentralization (i.e., where everyone in the organization is completely free to decide what they want to do or not do). As an example of such structural parameters, while employees of Minnesota Mining and Manufacturing (3M) are encouraged to take the initiative in suggesting new products and seeking support from others in the organization, new product teams must still meet specific financial measures at each stage of product development. Nonetheless, the real control is found in constant interaction among peers and the normative rules that develop informally among them.

It is not always necessary for an entire organization to be organic. Some units, such as research and development departments, may benefit from an organic structure because they face an unstable environment. Units that have a more stable environment, such as routine, administrative departments, may favour a mechanistic structure. Some units may borrow from both models. Customer service departments, for example, can build flexibility into responding to exceptional circumstances while maintaining standardized protocols for more typical situations.

The structures of organic organizations are informal, fluid, and constantly changing to identify and develop responses to new problems and opportunities. Authority and responsibility shifts from one situation to another. Groups are established, complete their work, and disband, and a single employee may belong to several temporary teams at the same time. In organic organizations there is diminished emphasis on superior-subordinate roles in favour of dispersed initiative. Roles, tasks, and responsibilities are not limited by rigid, vertical boundaries of hierarchy for decision-making, communication, coordination, and control. Relations and interactions between personnel and units continually change, and managers and other employees must figure out which relations and interactions will be most effective for each particular problem or opportunity.

SPECIFIC TYPES OF ORGANISATIONAL CHANGE

There are different overall types of organizational change, including planned versus unplanned, organization-wide versus change primarily to one part of the organization, incremental (slow, gradual change) versus transformational (radical, fundamental), etc.Knowing which types of change you are doing helps all participants to retain scope and perspective during the many complexities and frequent frustrations during change.

ORGANISATIONAL EXCELLENCE

Men and women who have held official positions of some importance start developing phobias when retirement is imminent and start looking for family support when they have discarded the self-same families and lived in a world of selfish indulgences when the going was good. People also seek conformity, (to norms convenient to their personal well-being), as an insurance against perceived depletion of power and privileges, which they have been enjoying, and deep down they realise that this feel-good experience may be short-lived. Executives who have lived high profile lives often fear that after retirement they may become socially insignificant. These are those who harboured feelings based on false consciousness. On the other hand persons who have lived their lives ethically do not harbour such fear since ethics sustains authority.

There are several examples like the ones cited above but from times immemorial men have fashioned unfreedom, which is a large measure of conformity, as a bribe for self-perpetuation. In the words of Peters and Waterman, men quite often willingly shackle themselves to mundane, pen pushing, clerical nine to five jobs, if only the cause is perceived to be something great. Management is as guilty of perpetuating this sense of false consciousness amongst its workforce as is the employee for lamely accepting it.

Organised ritual based religion performs the same function for the larger civil society as the government does for the over-staffed and under ambitious minions in the civil service. Little does the government realise what moral crime it is committing against humanity. How can, for instance, an employee or a person be rendered surplus? It means *either one or all* of these four things. The recruiter goofed up. The top management failed to create a competitive organisation. The manager was not creative enough to generate work. The Human Resources department failed to develop the employee so that he/she could add value to the organisation and thus not become surplus. How can organisational excellence be achieved under such conditions and why does the HR Expert not have the gumption to shoulder the blame?

Positive economists do not fail to remind us that all value is created by labour that capital man made aid to production that production is the creation of value and consumption is but the transformation or destruction thereof. Sociologists constantly harp on the theme that in the pursuit of a better quality of life mankind has evoked interest in the sciences, the arts, religion and the philosophies in the hope that their lot will be better. Historians vouchsafe that man has sought excellence through his efforts has left behind markings of his footsteps on the sands of time. And yet when we go to a symposium of experts on excellence all we hear is words like systems, processes, procedures and market competition. Where did man, the basic creator of value disappear in this new calculus? This short piece is aimed at finding out why Indian companies do not attain the much-needed maturity to be truly competitive and thereby approach excellence.

It is found that only a few top companies like Wipro, Infosys and TCS in information technology, Ranbaxy and Sun in the pharmaceutical industry, Tata Steel, Godrej Industries, Tata Motors, HDFC and ICICI Banks, NDDB, SAIL, BHEL and NTPC, have attained a measure of excellence. This is because in other organisations the basic labour force, the man behind the machine or the desk, is conveniently asked to play a secondary role to technology. Surely, the CEO must have gone to a good B-School or an elite University. How could he/she but not know this simple fact? Why is there such a great divide between the theory and practice of people management and why is Organisational Excellence so illusive? Is it enough to devise and put into place an ethical culture and good governance practices? What is the role of Human Resources Management in helping the organisation to achieve excellence? The fact that business ethics and corporate governance is an intrinsic part of building Organisational Excellence can no longer be doubted. But in exploring the reasons behind the great divide between the theory and practice of management and the failure to achieve excellence, the role of Human Resources Management (as a specialisation) is thrown up for critical examination. This is an aspect of objective social reality that this short paper shall try to examine briefly.

Themes that generally emerge when we examine the gap between the theory and practice of management unmistakably point towards the fact that the HR function does not or is not allowed to pull its weight. We shall address six of the main ones, which if satisfactorily approached even if not wholly resolved will make a positive difference. These are:

- The competitive challenges faced by organisations.
- The extent to which TQM (total quality management) has been practised.
- Evolution of a learning organisation.
- Outsourcing the legitimate HR functions.
- Self-assessment frameworks and their role in modern times.
- Approaches founded on performance management systems.

Competitive Challenges: Organisations have long since realised that the only way to thrive on the cutting edge of market competition is to use technology to convert their core competence into competitive advantage. Organisations consequently need to learn how best to respond and adapt to the plethora of discontinuities of life. An organisation is known by the type and competence of people it has and not by the physical infrastructure it exhibits. The service mindset has to pervade every facet of corporate life and organisations have to compete along several dimensions to meet customer needs.

Visionary leadership is not enough since it has to be sustainable in nature and this comes from it being value centred at the same time. Change orientation in every facet of organisational life must be such that divergent skills and behaviours combine to achieve convergent aims and objectives. Organisations must have goal focus, customer focus and market focus at the same time. The customer and the market have become more important that profits and technology. The latter were the drivers of progress in the 1980s and the 1990s while the former have become the organisational drivers in the 2000 era. The key question posed is thus: do organisational results match the demands and expectations of the market and the customer? If not, then sustainability will be short lived.

Total Quality Management: It is most unfortunate that TQM is a term that is heard only in classrooms and seminar halls. It is not as often seen in organisational practice, as it ought to be. The promised gains of the LPG exercise are slow in materialising and the ranks of the self-employed and the un-employed persons are rising. If one reads the daily newspapers one finds out that scheming and scamming have become the order of the day. Entrepreneurs get rich at the expense of the organisations they have promoted.

If one reads academic journals and the research findings that are reported in them, another story emerges. Top management is not visibly involved in integrating quality into strategic planning. Even managers who speak most vociferously about change and transparency are unable to change their own ways of doing things or accept that they need to be transparent themselves. How should senior managers provide the much needed direction and how should performance be measured at the strategic level remain issues of serious concern? The Eastern Hotels Group has a lesson or two to teach their industry in this arena.

Financial compulsions and short run profit motives overshadow the imperatives of quality management and customer satisfaction. There is a visible gap between knowing what out to be done and doing what needs to be done. Perhaps the reason is that top management is caught up in the glamour of event management and short run actual cost considerations that it is in market orientation and long run opportunity cost considerations.

Hence a sea change in the top-level mindset is called for and quality must become a way of life. Concepts like Economic Value added (EVA) and severe consequence management (SCM) need to be pursued with zeal and actualised. The corporate strategy itself must be well thought out, well managed and be capable of converting customers into clients.

The Learning Organisation: Top management needs to have the maturity and the ability to understand and learn from organisational performance and outcomes. New compulsions demand that new organisational competencies emerge and new strategies are harnessed. Value centred leadership, visionary stewardship, the ability to learn from the mistakes made by themselves and those of others, seizing opportunities for change, creativity and innovation, and restoring the dignity of labour are the new HR mantras of the 21st century. How these competencies are recognised, how the existing talent is nurtured, how to create multiple teams that complement each other's efforts and how to get ahead of oneself become the major HR concerns. Empowerment and accountability must go hand in hand. HR strategy for this reason must be inseparable from HR execution. The road ahead must be well mapped, realistic and user friendly if organisational goals are to be achieved.

People Management must replace the term Human Resource Management since people are not inanimate resources like silica sand and cement bricks. The sense of belonging must be so enhanced that people start saying and acting as if (a) the organisation belongs to me, and (b) I belong to the organisation. Hence new managerial competencies need to be developed and this has to be so demonstrated that people would "want" to be a part of the growth process. Above all else, management must learn that since it cannot alter the direction of the wind, it can always alter the angle of its sails to get ahead with its dream.

Outsourcing HR Functions: The bogey of Business Process Outsourcing that has come fast on the heels of Business Process Re-engineering seems to have caught the fascination of many a HR Chief. It first began with hiring persons equipped to fill out various returns required under Indian labour legislation.

Most of these persons were either moonlighting from government offices where they were employed full time or had been working as such prior to their retirement and knew how to pull the right ropes. The came the period when trainers were hired to impart special skills and behaviours that HR departments felt that their organisation's employees needed.

That scientific diagnosis often did not precede such activities is well known. Then came the era of hiring the services of placement consultants to pick and choose manpower and thereby cut down recruitment costs. At this point something nefarious happened. The trainers tied up (informally) with the recruiting consultants. The level of attrition rose especially in the software industries sector and together the financial fortunes of the trainer and the recruitment consultant flourished. Just about the same time (in the 1990s) companies went in for Business Process Re-engineering. The task was invariably assigned to managerial greenhorns and not to the experienced people managers. The result was that organisational structures were flattened and accountability was emphasised but delegation of authority did not follow. Just when top management decided that BPR was not such a good thing after all, the new mantra of BPO shot into prominence. The more enlightened companies introduced computer packages like Peoplesoft and Ramco while the not so bright management actually outsourced the HRIS function. Incessant pressures on organisations to be goal focused, customer focused and market focused remained. Retention, reward & recognition and career development became a part of HR strategy. Structures and functions concurrently collapsed just as organisational goals and managerial roles began to unfold and expand. It was then that Strategic HR consciously replaced HR Strategy, and People Management as a specialisation, reached its maturity. *Experience* which was a treasured trait until fairly recently was replaced by *talent* which was a trait that was prized and hence managed and developed. As has been argued in *Geometry of HR* (2002), one thing that cannot be outsourced is Corporate Culture Building. This is what People Management today must be concerned with and this is where ethics and governance have a positive role to play.

Self-Assessment Frameworks: In the journey towards excellence, self-assessment awards and frameworks lay claim to being the vehicles for organisational learning. Unfortunately the weight of evidence is against it and more organisations use the business excellence model than those who use self-assessment frameworks. But the business excellence model itself is a questionable construct in several organisations. This is because of many reasons. Firstly the validity of Business Results criteria is open to question. How can star organisations that were awarded accolades for good governance become sick babies the very next year, as was the case with UTI and L&T? Secondly the measurement of the link between internal improvements (called enablers) and business results is often blurred. Language used in the business excellence model is often open to varied interpretations, unless of course it happens to be written in Sanskrit where one word can have only one meaning! User judgement in identifying key processes is not always uniform and unless a high degree of quality maturity has been attained, award frameworks will remain public relations exercises that use political leverage to build corporate brands.

We often hear of great scientists who have almost become cult figures in the 2000 decade, when in reality their last real research paper was written three decades ago. Today they may well have become social icons who are out of touch with the very subject in which they are held to be gurus. If competitiveness must be sustained then it follows that excellence must become a habit and all managerial activity in an organisation must be value-centred. One must constantly compete with oneself and erase the word impossible from one's lexicon. It was once believed that the human body was incapable of running the mile in less than 4 minutes. When Roger Banister broke the illusion clocking 3 minutes and 59 seconds in May 1954 people took it as a freak incident. But how can we also thing so since the 4 minute a mile record has been broken over a few hundred times since then? Hence self-assessment is limited by the fact that one wishes to improve constantly and realises that the last rung of the ladder risen to is not to be rested upon but used to climb higher still. The answer therefore lies in developing multi-dimensional and dynamic models for monitoring corporate performance.

ORGANIZING DECISIONS

When organizing, managers must make decisions about the division of labour and work specialization, departmentalization, chain of command, span of management, centralization, and formalization. Collectively, these decisions are often called organizational design.

Division of Labour or Specialization

More than two centuries ago Adam Smith concluded that division of labour contributes to increased productivity and efficiency by allowing workers to specialize and become proficient at a specific task. This principle, coupled with technological advances, made possible the tremendous productivity of industrial companies during most of the twentieth century. By the 1940s most manufacturing jobs in developed nations were highly specialized, with workers performing specific, standardized, and repetitive tasks. This resulted in reduced staffing, training, and compensation costs, since highly skilled workers were often not necessary. In addition, since employees were doing the same task repetitively, they tended to become very good at it.

Despite the improvements in productivity made possible by the division of labour, managers must be aware of the negative aspects of specialization: fatigue, stress, boredom, low quality products, absenteeism, and turnover. Such problems have led to programs geared toward job enlargement and job enrichment.

Departmentalization

After the work to be completed is organized into identifiable jobs through a process of dividing labour, jobs are then combined into logical sections or departments. Doing so allows for effective coordination of effort. There are many ways to departmentalize, each of which has important advantages and disadvantages. One of the most common forms is functional departmentalization, which involves grouping similar jobs into a common department, such as accounting, sales, human resources, and engineering. Another form is product departmentalization, which involves organizing around an enterprise's various product lines. Other ways of departmentalizing include organizing by customer and by geographic territory. In practice, most large companies use a hybrid form of departmentalization, which means they combine one or more of the above methods to form their organizational structure.

Chain of Command

The chain of command is a line of authority extending from the top to the bottom of the organizational structure. Classic principles of organizing emphasize that one must be aware of the need to define the extent of managers' responsibility and authority by specifying their place in the chain of command. Another principle of organizing related to the chain of command is called the unity of command, which states that a person should have only one superior to whom he or she must report.

Span of Management

The span of management, often called the span of control, is the number of individuals who are directly responsible to a particular manager. A classic principle of organizing suggests that there are definite limits to the number of subordinates one manager can supervise effectively. When organizing, managers must keep these limits in mind. Wide spans of management lead to flatter organizational structures with fewer layers of management, and are thus considered more efficient. However, if spans become too wide managers may not be able to provide adequate direction to subordinates. Narrow spans of management lead to tall organizational structures with many layers of management. Although narrower spans of management allow for closer supervision of subordinates they have many drawbacks, including cost, communication problems, and difficulty in developing the initiative and autonomy of subordinates.

In general, the trend is toward wider spans of management, with an accompanying decrease in management hierarchy. Technological advances in information processing and communication have made wider spans of management more feasible.

Degree of Centralization

Another organizing decision is the degree of centralization in the organizational structure. If decision-making authority in an organization is highly centralized, then most major decisions are made at the upper levels of the structure. Conversely, if decision-making authority is decentralized, important decisions are often made at lower levels of the hierarchy. The degree of centralization that is appropriate for a given organization depends upon many factors, including the nature of the environmental conditions that face the enterprise, the characteristics and abilities of lower-level employees, and the size of the enterprise. Many organizations are favoring a greater degree of decentralization of their decision-making authority.

Formalization

The degree of formalization in an enterprise refers to the degree to which there are standardized rules and procedures governing the activities of employees. A company with a high degree of formalization is characterized by detailed job descriptions and clearly defined policies and procedures covering a wide variety of employee behaviors. Conversely, a company with a low level of formalization is characterized by non-structured jobs and fewer explicit policies and procedures. As companies grow larger, a certain amount of formalization is inevitable. Employees require some direction in their job responsibilities and in the procedures required for consistency within the organization's production schema. When organizing, however, managers should be aware of the costs of excessive formalization, which may include stifling employee creativity and innovation as well as slowing the organization's responsiveness to critical issues and problems.

FACTORS AFFECTING ORGANIZING DECISIONS

There is no standard formula for the best way to organize an enterprise. Several factors have been shown to influence organizing decisions. Among the most important of these factors are strategy, size, environmental conditions, and technology.

Managers organize in order to achieve the objectives of the enterprise for which they work. Thus, the strategy of the enterprise affects organizing decisions. Changes in strategy frequently necessitate changes in the way the enterprise is organized.

Small enterprises tend to exhibit less formalization, centralization, and complexity in their organizational structure. Nevertheless, enterprises of the same size may be organized quite differently because of differences in strategy, environmental conditions, and technology.

Environmental Conditions

The key factor in the external environment that is relevant to organizing is uncertainty. Some enterprises face competitive environments that change rapidly and are quite complex, while others face relatively stable conditions. Generally, turbulent environments call for organizing decisions that lead to less formalization and centralization in the organizational structure.

Technology

The processes by which an enterprise transforms inputs into outputs may also affect organizing decisions. Some research suggests that organizing decisions that lead to high degrees of formalization, centralization, and work specialization are more appropriate for routine technologies and that the converse is true for non-routine technologies.

3

Planning Research Strategy

STRATEGIC PLANNING

Strategic planning is an organization's process of defining its strategy, or direction, and making decisions on allocating its resources to pursue this strategy, including its capital and people.

Various business analysis techniques can be used in strategic planning, including SWOT analysis and PEST analysis. The outcome is normally a strategic plan which is used as guidance to define functional and divisional plans, including Technology, Marketing, etc. Strategic Planning is the formal consideration of an organization's future course. All strategic planning deals with at least one of three key questions:

1. "What do we do?"
2. "For whom do we do it?"
3. "How do we excel?"

In business strategic planning, the third question is better phrased "How can we beat or avoid competition?" In many organizations, this is viewed as a process for determining where an organization is going over the next year or more-typically 3 to 5 years, although some extend their vision to 20 years. In order to determine where it is going, the organization needs to know exactly where it stands, then determines where it wants to go and how it will get there.

The resulting document is called the "strategic plan". It is also true that strategic planning may be a tool for effectively plotting the direction of a company; however, strategic planning itself cannot foretell exactly how the market will evolve and what issues will surface in the coming days in order to plan your organizational strategy. Therefore, strategic innovation and tinkering with the 'strategic plan' have to be a cornerstone strategy for an organization to survive the turbulent business climate. (Ziaur Rahman, IITM, iitmbd.org, Dhaka)

Vision, Mission and Values

Vision: Defines where the organization wants to be in the future. It reflects the optimistic view of the organization's future.

Mission: Defines where the organization is going now, basically describing the purpose, why this organization exists.

Values: Main values protected by the organization during the progression, reflecting the organization's culture and priorities. Strategic planning saves time, every minute spent in planning saves ten minutes in execution.

The purpose of individual strategic planning is for you to increase your return on energy, the return on the mental, emotional, physical and spiritual capital you have invested in your life and career.

Every minute an individual spends planning their goals, activities and time in advance saves ten minutes of work in the execution of those plans. Careful advance planning gives you a return of ten times, or 1,000%, on your investment of mental, emotional and physical energy. (The 100 Absolutely Unbreakable Laws of Business Success.)

Methodologies

There are many approaches to strategic planning but typically a three-step process may be used:

- Situation-evaluate the current situation and how it came about.
- Target-define goals and/or objectives (sometimes called ideal state)
- Path-map a possible route to the goals/objectives

One alternative approach is called Draw-See-Think

- Draw-what is the ideal image or the desired end state?
- See-what is today's situation? What is the gap from ideal and why?
- Think-what specific actions must be taken to close the gap between today's situation and the ideal state?
- Plan-what resources are required to execute the activities?

An alternative to the Draw-See-Think approach is called See-Think-Draw

- See-what is today's situation?
- Think-define goals/objectives
- Draw-map a route to achieving the goals/objectives

In other terms strategic planning can be as follows:

- Vision-Define the vision and set a mission statement with hierarchy of goals
- SWOT-According to the desired goals conduct analysis
- Formulate-Formulate actions and processes to be taken to attain these goals

- Implement-Implementation of the agreed upon processes
- Control-Monitor and get feedback from implemented processes to fully control the operation

Situational Analysis

When developing strategies, analysis of the organization and its environment as it is at the moment and how it may develop in the future, is important. The analysis has to be executed at an internal level as well as an external level to identify all opportunities and threats of the new strategy. There are several factors to assess in the external situation analysis:

1. Markets (customers)
2. Competition
3. Technology
4. Supplier markets
5. Labor markets
6. The economy
7. The regulatory environment

It is rare to find all seven of these factors having critical importance. It is also uncommon to find that the first two-markets and competition-are not of critical importance. (Bradford "External Situation-What to Consider")

Analysis of the external environment normally focuses on the customer. Management should be visionary in formulating customer strategy, and should do so by thinking about market environment shifts, how these could impact customer sets, and whether those customer sets are the ones the company wishes to serve. Analysis of the competitive environment is also performed, many times based on the framework suggested by Michael Porter.

Goals, Objectives and Targets

Strategic planning is a very important business activity. It is also important in the public sector areas such as education. It is practiced widely informally and formally. Strategic planning and decision processes should end with objectives and a roadmap of ways to achieve those objectives.

The following terms have been used in Strategic Planning: desired end states, plans, policies, goals, objectives, strategies, tactics and actions. Definitions vary, overlap and fail to achieve clarity. The most common of these concepts are specific, time bound statements of intended future results and general and continuing statements of intended future results, which most models refer to as either goals or objectives (sometimes interchangeably).

One model of organizing objectives uses hierarchies. The items listed above may be organized in a hierarchy of means and ends and numbered as follows: Top Rank Objective (TRO), Second Rank Objective, Third Rank Objective, etc. From any rank, the objective in a lower rank answers to the question "How?" and the objective in a higher rank answers to the question "Why?" The exception is the Top Rank Objective (TRO): there is no answer to the "Why?" question. That is how the TRO is defined.

People typically have several goals at the same time. "Goal congruency" refers to how well the goals combine with each other. Does goal A appear compatible with goal B? Do they fit together to form a unified strategy? "Goal hierarchy" consists of the nesting of one or more goals within other goal(s).

One approach recommends having short-term goals, medium-term goals, and long-term goals. In this model, one can expect to attain short-term goals fairly easily: they stand just slightly above one's reach. At the other extreme, long-term goals appear very difficult, almost impossible to attain. Strategic management jargon sometimes refers to "Big Hairy Audacious Goals" (BHAGs) in this context.) Using one goal as a stepping-stone to the next involves goal sequencing.

A person or group starts by attaining the easy short-term goals, then steps up to the medium-term, then to the long-term goals. Goal sequencing can create a "goal stairway". In an organizational setting, the organization may co-ordinate goals so that they do not conflict with each other. The goals of one part of the organization should mesh compatibly with those of other parts of the organization.

Mission Statements and Vision Statements

Organizations sometimes summarize goals and objectives into a mission statement and/or a vision statement:

While the existence of a shared mission is extremely useful, many strategy specialists question the requirement for a written mission statement. However, there are many models of strategic planning that start with mission statements, so it is useful to examine them here.

- *A Mission statement*: tells you what the company is now. It concentrates on present; it defines the customer(s), critical processes and it informs you about the desired level of performance.
- *A Vision statement*: outlines what a company wants to be. It concentrates on future; it is a source of inspiration; it provides clear decision-making criteria.

Many people mistake vision statement for mission statement. The Vision describes a future identity and the Mission describes why it will be achieved. A Mission statement defines the purpose or broader goal for being in existence or in the business. It serves as an ongoing guide without time frame. The mission can remain the same for decades if crafted well. Vision is more specific in terms of objective and future state. Vision is related to some form of achievement if successful.

A mission statement can resemble a vision statement in a few companies, but that can be a grave mistake. It can confuse people. The vision statement can galvanize the people to achieve defined objectives, even if they are stretch objectives, provided the vision is SMART (Specific, Measurable, Achievable, Relevant and Timebound). A mission statement provides a path to realize the vision in line with its values. These statements have a direct bearing on the bottomline and success of the organization.

Which comes first? The mission statement or the vision statement? That depends. If you have a new start up business, new program or plan to re engineer your current services, then the vision will guide the mission statement and the rest of the strategic plan. If you have an established business where the mission is established, then many times, the mission guides the vision statement and the rest of the strategic plan. Either way, you need to know where you are, your current resources, your current obstacles, and where you want to go-the vision for the future. Features of an effective vision statement may include:

- Clarity and lack of ambiguity
- Paint a vivid and clear picture, not ambiguous
- Describing a bright future (hope)
- Memorable and engaging expression
- Realistic aspirations, achievable
- Alignment with organizational values and culture, Rational
- Time bound if it talks of achieving any goal or objective.

To become really effective, an organizational vision statement must (the theory states) become assimilated into the organization's culture.

Leaders have the responsibility of communicating the vision regularly, creating narratives that illustrate the vision, acting as role-models by embodying the vision, creating short-term objectives compatible with the vision, and encouraging others to craft their own personal vision compatible with the organization's overall vision.

PRODUCTION PLANNING

Manufacturing of goods is a complicated process, specially in a big concern. The production planning manager has to think about advance planning of the production of goods very carefully. He has to see as to what is to be produced, how it is to be produced, and by what time should it be produced. It may be said that "production planning is a series of related and co-ordinated activities performed by not one but a number of different departmental groups, each activity being designed to systematize in advance the manufacturing efforts in its area." The production planning manager has to consult several persons from different departments and has to collect information about the goods to be produced, the quantity of the goods to be produced, the quality desired, the date when delivery is to be made to the customers, etc.

On the basis of all these informations he prepares a schedule. He has to see in advance the manpower available, the materials, the machines and money required for a pre-determined output in a given period of time.

In the words of S. N. Chary, "Production planning, in particular, would therefore consist mainly of the evaluation and determination of production inputs such as labour (manpower), machinery and equipment, materials, and utilities to achieve the desired goals. Definition of the goals is also of course, a part of production planning process."

Production planning should fulfil certain aims. The various aims of planning are:

(a) to produce greater volume of goods at the minimum possible cost;
(b) to facilitate production of goods in the shortest possible time;
(c) to capture a desired share of the market demand;
(d) to operate the plant at a pre-determined level of efficiency:
(e) to utilise to the maximum the available plant facilities;
(f) to create specified number of jobs;
(g) to produce goods with greater certainty;
(h) to bring a prescribed level of profit.

Production planning should, as far as possible, be flexible enough. The demand for the product may undergo a change with the change in the market situation and there should be enough scope for bringing about the changes in the plan.

In the absence of this flexibility the concern may have to face several problems. Such flexibility may be brought about by making changes in the inventory, or by varying the length of the work schedules and through fluctuating employment.

PRODUCTION CONTROL

Mere advance planning for production may not achieve the ends unless there is proper control over all the processes of production. Like other controls, it may mean only putting restraints, curbs, checks and other regulations but it is not so. As a matter of fact it also stimulates, activates and galvanizes. According to S.N. Chary, "Production control is the updating and revising procedure where, according to the requirements of implementation, the labour assignments, the machine assignments, the job priorities, the line speeds, the production routes etc. may be revised. It is basically a correcting mechanism which goes on throughout the implementation process of already drawn up production plan and schedule."

It regulates the orderly flow of materials in the manufacturing process from the raw state to the finished product. It alms to produce the right product, in the right quantity, of the right quality, at the right time, and by the best and least costly methods.

"Good production-control procedure means less work in process, decreased stock inventories, and more rapid turnover, which in turn results in less capital tied up in idle material and greater earnings on money thus invested." As such it is not simply regulatory but facilitating and coordinating too. It should, however, be noted that it adds nothing to the value of the product, with the possible exception of additional time and place utility. Production control involves both planning and control. It involves advance planning necessary to achieve an orderly flow of materials, and the execution of that plan to ensure that the desired control is maintained. Planning without control cannot fulfil its aims and control may not be successful without planning.

Steps in Production Control

To bring about the desired results on the basis of advance planning of production several steps may have to be taken by way of control. In practice, we find that four main steps are taken in production control. They are: routing, scheduling, dispatching, and inspection.

Routing

Production routing determines the way in which the goods have to be produced. It fundamentally involves establishment of the operations, their path and sequence, and proper class of machines and personnel required for these operations. Once the layout for production has been determined the routing involves the determination of path or sequence that item should follow as the goods are in the process of being produced. It should be kept in mind that routing should be determined in such a manner that there is minimum handling and back-tracking of the materials in the process of production.

Routing is not a serious problem in case of a continuous manufacturing plant with product layout. Under this process, the standardised goods are produced in the standardised way and there is repetitive And continuous process of production; therefore routing once determined and established would have simply to be repeated. But the problem is not so simple in case of job-order production and batch production with process layout. It would involve determination of the routing of the materials every time a product is to be manufactured because the sequence of production may not have to be repeated again and again as in the earlier case. In such a case, there may be several ways for routing materials. Then, there may be need for the determination of the best routing. Through experience and experimentation the best routing may be obtained and a route card should be prepared outlining the whole path and sequence and the proper class of machines and personnel required for a particular operation. This should be done in planning section.

The routing procedure for a new product or a new part may consist of as many as seven separate activities. They are:

(i) analysis of the article to determine what to make and when to purchase;
(ii) analysis of the article to determine what materials are needed;
(iii) determination of the manufacturing operations and their sequence;
(iv) determination of lot sizes;
(v) determination of scrap factors;
(vi) analysis of the cost of the article; and
(vii) organisation of production control forms.

Several considerations affect the routing procedures. The type of manufacturing employed and the manner in which the plant is laid out have their impact upon it. It is affected by the individual characteristic of the physical equipment. The machines may differ in age, condition, degree of precision, capacity, range, speed tooling, and a host of other qualities. They are also affected by the availability.of physical plant equipment. When the cheapest machine or method cannot be utilised because of machine-load conditions, breakdowns etc., alternative routing should be prepared. Similarly, due consideration should be given to the human factors involved in the process. It is to be noted that the "operators vary as to experience and such capability characteristics as visual acuity, manual dexterity, physical limitations, intelligence, and emotional stability". Under such circumstances, this factor cannot be ignored while determining the routing procedures.

Work Scheduling

Work scheduling involves determining when work should start and how much work should be completed during a certain period of time. In other words, "Scheduling is that phase of production control which rates the work in the order of its priority and then provides for its release to the plant at the proper time and in the correct sequence. Thus scheduling is concerned with when work shall be performed on a product part." It is undoubtedly an important part of production control as all future process of production is based on it. It is by scheduling that the entire groundwork is laid for all subsequent steps.

How much time or how many hours it will take to manufacture each component part of the product is an important factor which must be given due weighted in production planning and control. It must be known before hand as to what parts are to be produced, in what sequence they have to be produced and within what period they must be completed. It is scheduling which serves these purposes. In practice, scheduling starts with the preparation of the master schedule. Such schedule is prepared on the basis of the past experience and consultations with related departments. It may be weekly, monthly or even annual. The form in which it is prepared may vary from concern to concern. A simplified form of master schedule has been given below: Preparation of master schedule helps control over-production to a very great extent. It provides a suitable and convenient means of keeping a running total of the production requirements. It also facilitates the production manager to undertake advance planning for any shifting of emphasis from one product to another or increase or decrease in production requirements in the changed situation. It helps in controlling as it furnishes data for calculating the backlog of work. With the help of such schedule, a possible or definite date of delivery of goods may be given to the customer.

Production Dispatching

It is one of the two 'action elements' of the production control, the other being inspection. It is the function which initiates the production as planned, routed and scheduled. It involves actual granting of permission to proceed according to plans already laid down. In general, by dispatching we mean starting something on its way, but in connection with manufacturing it indicates "the method or process of assigning work to define work places, and where necessary, accelerating or retarding its progress from point to point." It also involves issuance of orders in terms of their priority as determined by scheduling. Dispatching, in general, involves the following activities:

(i) The assignment of work to the machines or workplaces.

(ii) The authority to prepare, assemble, and issue to their point of use the necessary materials, tools, fixtures, and gauges.

(iii) The issuance of orders and production forms necessary to the performance of work and to the reporting of production, payroll, and cost data,

(iv) The responsibility for controlling the progress of material at each operation; for making the necessary adjustments to schedule and work assignments as conditions change or as unpredictable emergencies occur.

(v) The authority to move work from one operation to the next.

(vi) The liaison function linking the plans of the routing and scheduling office with the performance of the manufacturing divisions.

The dispatching procedures may be centralised or decentralised, depending upon the size and nature of the concern and the nature of commodity to be produced. A centralised dispatching involves the dispatching of orders from the central dispatching division directly to the machine or work station, whereas a decentralising dispatching procedure involves "issuing manufacturing schedules or work orders in blanket fashion to the foreman or "dispatch clerk within each department who must then determine the relative sequence in which those orders will be started within that department."

Follow-up and Inspection

The follow-up or expediting is that part of the production control which is concerned with regulation of the progress of materials and parts through the production process. It has to be seen that production is being carried on in accordance with the plan laid out and there is no lack of co-ordination. If there is any variation from the plan, that has to be corrected properly. If the goods have to be produced as planned, it is necessary that proper follow-up or expediting must be undertaken even though it may involve a little cost. Such cost may have to be incurred if any expediter is appointed. He may have to maintain proper records about the progress made; delays, if any. In the supply of materials, tools etc.; the performance as compared to the plan; and proper coordination.

He learns about the delays through the analysis of the production reports and through personal observations. It would be better if he anticipates such delays in advance and make efforts to prevent them before they actually develop. As it is well-known, an ounce of prevention is far more valuable than a pound of cure where delays in industrial production are concerned.

Similarly, inspection is needed to ensure the quality of the goods produced. This has to be checked up while the goods are in the process of production, inspection of the quality of finished material is necessary but at the same time it is highly necessary in the process of production. Because once the goods have been produced, any correction is not possible and the rejection of goods on a large scale would involve high cost to the concern. Therefore, it is necessary to carry on inspection from the stage of the procurement of raw material to that of the actual finishing of the material. Thus, it would involve checking the quality of raw material, the tools and machines available, the skill and efficiency of workers who are looking after the production, actual performance in each stage or process of production. It is continuous inspection of the production of the goods which matters. To carry on such inspection, sample checking may be resorted to and the whole lot need not be inspected unless there is some special reason. Inspection is one of the important tools of production control, but it has certain limitations:

(a) it cannot indicate the extent to which the quality of a product, requiring high degree of precision, conforms to the pre-determined standards;
(b) it does not give all the information required for rejection of substandard quality goods produced, specially when the standardised goods are produced on a large scale; and
(c) it does not prevent the scrapping of the finished goods on a large scale which would involve huge cost to the concern.

Because of these limitations of inspection in ensuring quality control, statistical quality control method is being adopted by the industrial concerns.

MARKETING STRATEGY : PLANS AND PLANNING

Plans are needed to clarify what kinds of strategic objectives an organization would like to achieve and how this is to be done. Such plans must consider the amount of resources available. One critical resource is capital.

Microsoft keeps a great deal of cash on hand to be able to "jump" on opportunities that come about. Small startup software firms, on the other hand, may have limited cash on hand. This means that they may have to forego what would have been a good investment because they do not have the cash to invest and cannot find a way to raise the capital.

Other resources that affect what a firm may be able to achieve include factors such as:

- *Trademarks/brand names*: It would be very difficult to compete against Coke and Pepsi in the cola market.
- *Patents*: It would be difficult to compete against Intel and AMD in the microprocessor market since both these firms have a number of patents that it is difficult to get around.
- *People*: Even with all of Microsoft's money available, it could not immediately hire the people needed to manufacture computer chips.
- *Distribution*: Stores have space for only a fraction of the products they are offered, so they must turn many away. A firm that does not have an established relationship with stores will be at a disadvantage in trying to introduce a new product.

Plans are subject to the choices and policies that the organization has made.

Some firms have goals of social responsibility, for example. Some firms are willing to take a greater risk, which may result in a very large payoff but also involve the risk of a large loss, than others. Strategic marketing is best seen as an *ongoing and never-ending process*.

Typically:

- The organization will identify the *objectives* it wishes to achieve. This could involve profitability directly, but often profitability is a long term goal that may require some intermediate steps. The firm may seek to increase market share, achieve distribution in more outlets, have sales grow by a certain percentage, or have consumers evaluate the product more favourably. Some organizations have objectives that are not focused on monetary profit—*e.g.*, promoting literacy or preventing breast cancer.
- An analysis is made, taking into consideration issues such as *organizational resources, competitors, the competitors' strengths, different types of customers, changes in the market*, or *the impact of new technology*.
- Based on this analysis, a plan is made based on *tradeoffs between the advantages and disadvantages* of different options available.
- This strategy is then carried out. The firm may design new products, revamp its advertising strategy, invest in getting more stores to carry the product, or decide to focus on a new customer segment.
- After implementation, the results or outcome are evaluated. If results are not as desired, a change may have to be made to the strategy. Even if results are satisfactory, the firm still needs to monitor the environment for changes.

Levels of Planning and Strategies

Plans for a firm can be made at several different levels. At the *corporate* level, the management considers the objectives of the *firm as a whole*. For example, Microsoft may want seek to grow by providing high quality software, hardware, and services to consumers. To achieve this goal, the firm may be willing to invest aggressively. Plans can also be made at the *business unit* level. For example, although Microsoft is best known for its operating systems and applications software, the firm also provides Internet access and makes video games. Different managers will have responsibilities for different areas, and goals may best be made by those closest to the business area being considered. It is also more practical to hold managers accountable for performance if the plan is being made at a more specific level. Boeing has both commercial aircraft and defence divisions.

Each is run by different managers, although there is some overlap in technology between the two. Therefore, plans are needed both at the corporate and at the business levels. Occasionally, plans will be made at the *functional* level, to allow managers to specialize and to increase managerial accountability.

Marketing, for example, may be charged with increasing awareness of Microsoft game consoles to 55% of the U.S. population or to increase the number of units of Microsoft Office sold. Finance may be charged with raising a given amount of capital at a given cost. Manufacturing may be charged with decreasing production costs by 5%. The firm needs to identify the business it is in. Here, a balance must be made so that the firm's scope is not defined too narrowly or too broadly. A firm may define its goal very narrowly and then miss opportunities in the market place.

For example, if Dell were to define itself only as a computer company, it might miss an opportunity to branch into PDAs or Internet service. Thus, they might instead define themselves as a provider of "information solutions." A company should not define itself too broadly, however, since this may result in loss of focus. For example, a manufacturer of baking soda should probably not see itself as a manufacturer of all types of chemicals. Sometimes, companies can define themselves in terms of a customer need. For example, 3M sees itself as being in the business of making products whose surfaces are bonded together. This accounts for both Post-It notes and computer disks. A firm's mission should generally include a discussion of the *customers served*, the kind of technology involved, and the markets served. Several issues are involved in selecting target customers. As suggested, consider these in more detail within the context of segmentation, but for now, the firm needs to consider issues such as:

- The *size* of various market segments;
- How well these segments are being *served by existing firms*;
- *Changes* in the market—*e.g.*, growth of segments or change in technology;
- How the firm should be *positioned*, or seen by customers. For example, Wal-Mart positions itself as providing value in retailing, while Nordstrom's defines itself more in terms of high levels of customer service.

The Boston Consulting Group matrix provides a firm an opportunity to assess *how well its business units work together*. Each business unit is evaluated in terms of two factors: *market share* and the *growth prospects* in the market. Generally, the larger a firm's share, the stronger its position, and the greater the growth in a market, the better future possibilities.

Four combinations emerge:

- A *star:* Represents a business unit that has a high share in a growing market. For example, Motorola has a large share in the rapidly growing market for cellular phones.
- A *question mark:* Results when a unit has a small share in a rapidly growing market. The firm's position, then, is not as strong as it would

have been had its market share been greater, but there is an opportunity to grow. For example, Hewlett-Packard has a small share of the digital camera market, but this is a very rapidly growing market.

- A *cash cow:* Results when a firm has a large share in a market that is not growing, and may even be shrinking. Brother has a large share of the typewriter market.
- A *dog:* Results when a business unit has a small share in a market that is not growing. This is generally a somewhat unattractive situation, although dogs can still be profitable in the short run. For example, Smith Corona how has a small share of the typewriter market.

Firms are usually best of with a portfolio that has a *balance* of firms in each category. The cash cows tend to *generate cash but require little future investment*. On the other hand, stars generate some cash, but *even more cash is needed to invest in the future*—for research and development, marketing campaigns, and building new manufacturing facilities. Therefore, a firm may take excess cash from the cash cow and divert it to the star.

For example, Brother could "harvest" its profits from typewriters and invest this in the unit making colour laser printers, which will need the cash to grow. If a firm has cash cows that generate a lot of cash, this may be used to try to improve the market share of a question mark. A firm that has a number of promising stars in its portfolio may be in serious trouble if it does not have any cash cows to support it. If it is about to run out of cash—regardless of how profitable it is— is becomes vulnerable as a takeover target from a firm that has the cash to continue running it. A SWOT analysis is used to help the firm *identify effective strategies*.

Successful firms such as Microsoft have certain strengths. Microsoft, for example, has a great deal of technology, a huge staff of very talented engineers, a great deal of experience in designing software, a very large market share, a well respected brand name, and a great deal of cash. Microsoft also has some weaknesses, however: The game console and MSN units are currently running at a loss, and MSN has been unable to achieve desired levels of growth. Firms may face opportunities in the current market. Microsoft, for example, may have the opportunity to take advantage of its brand name to enter into the hardware market. Microsoft may also become a trusted source of consumer services. Microsoft currently faces several threats, including the weak economy. Because fewer new computers are bough during a recession, fewer operating systems and software packages. Rather than merely listing strengths, weaknesses, opportunities, and threats, a SWOT analysis should suggest how the firm may use its strengths and opportunities to overcome weaknesses and threats. Decisions should also be made as to how resources should be allocated. For example, Microsoft could either decide to put more resources into MSN or to abandon this unit entirely. Microsoft has a great deal of cash ready to spend, so the option to put resources towards MSN is available. Microsoft will also need to see how threats can be addressed. The firm can earn political good will by engaging in charitable acts, which it has money available to fund. For example, Microsoft has donated software and computers to schools. It can forego temporary profits by reducing prices temporarily to increase demand, or can "hold out" by maintaining current prices while not selling as many units.

CONCEPT OF MARKETING

After World War II, the variety of products increased and hard selling no longer could be relied upon to generate sales. With increased discretionary income, customers could afford to be selective and buy only those products that precisely met their changing needs, and these needs were not immediately obvious. The key questions became:

- What do customers want?
- Can we develop it while they still want it?
- How can we keep our customers satisfied?

In response to these discerning customers, firms began to adopt the *marketing concept*, which involves:

- Focusing on customer needs before developing the product
- Aligning all functions of the company to focus on those needs
- Realizing a profit by successfully satisfying customer needs over the long-term

When firms first began to adopt the marketing concept, they typically set up separate marketing departments whose objective it was to satisfy customer needs. Often these departments were sales departments with expanded responsibilities. While this expanded sales department structure can be found in some companies today, many firms have structured themselves into marketing organizations having a company-wide customer focus. Since the entire organization exists to satisfy customer needs, nobody can neglect a customer issue by declaring it a "marketing problem"-everybody must be concerned with customer satisfaction. The marketing concept relies upon marketing research to define market segments, their size, and their needs. To satisfy those needs, the marketing team makes decisions about the controllable parameters of the marketing mix.

MARKETING BASICS

What exactly is marketing and why is it important to you as an entrepreneur? Simply stated, marketing is everything you do to place your product or service in the hands of potential customers. It includes diverse disciplines like sales, public relations, pricing, packaging, and distribution. In order to distinguish marketing from other related professional services, S.H. Simmons, author and humorist, relates this anecdote. "If a young man tells his date she's intelligent, looks lovely, and is a great conversationalist, he's saying the right things to the right person and that's marketing. If the young man tells his date how handsome, smart and successful he is — that's advertising. If someone else tells the young woman how handsome, smart and successful her date is — that's public relations."

You might think of marketing this way. If business is all about people and money and the art of persuading one to part from the other, then marketing is all about finding the right people to persuade. Marketing is your strategy for allocating resources (time and money) in order to achieve your objectives (a fair profit for supplying a good product or service).

Yet the most brilliant strategy won't help you earn a profit or achieve your wildest dreams if it isn't built around your potential customers. A strategy that isn't based on customers is rather like a man who knows a thousand ways to make love to a woman, but doesn't know any women. Great in theory but unrewarding in practice. If you fit the classic definition of an entrepreneur (someone with a great idea who's under-capitalized), you may think marketing is something you do later — after the product is developed, manufactured, or ready to sell. Though it may feel counter-intuitive, marketing doesn't begin with a great idea or a unique product. It begins with customers — those people who want or need your product and will actually buy it. Entrepreneurs are in love with their ideas, and they should be. After all, why would anyone commit their energy, life savings, and no small part of their sanity to anything less than a consuming passion. Because entrepreneurs are passionate about their idea, product, or service, they innocently assume other people will feel the same. Here's the bad news — it just doesn't work that way!

People have their own unique perceptions of the world based on their belief system. The most innovative ideas, the greatest products, or a superior service succeed only when you market within the context of people's perceptions.

Context can be many things, singly or simultaneously. To name a few, you may market to your customers within the context of their wants, needs, problems solved, or situation improved. Entrepreneurs need to be aware of many other contexts, such as social and economic trends or governmental regulations, which we'll discuss another time.

People don't just "buy" a product. They "buy" the concept of what that product will do for them, or help them do for themselves. People who are overweight don't join a franchise diet centre to eat pre-packaged micro-meals.

They "buy" the concept of a new, thin, happy and successful self. Before you become consumed with entrepreneurial zeal and invest your life savings in a new venture, become a smart marketer. Take time at the beginning to discover who your potential customers are, and how to effectively reach them.

Without a plan, your entrepreneurial dream is really wishful thinking. While a marketing plan can be a map for success, remember that the map is not the territory. A strategy that ignores the customer isn't an accurate reflection of the landscape.

A good marketing plan can help you focus your energy and resources. But a plan created in a vacuum, based solely on your perceptions, does not advance the agenda. That's why market research, however simple or sophisticated, is important. Just keep in mind that research attempts to predict the future by studying the past. It reveals what people have done, and extrapolates what people *might* do — not what people *will* do. Planning is imperative, research is important, but there's no substitute for entrepreneurial insight. After all, as Mark Twain wrote, "You cannot depend on your eyes when your imagination is out of focus".

MARKET DEFINITION

In marketing, the term *market* refers to the group of consumers or organizations that is interested in the product, has the resources to purchase the product, and is permitted by law and other regulations to acquire the product. The market definition begins with the total population and progressively narrows as shown in the following diagram.

Beginning with the total population, various terms are used to describe the market based on the level of narrowing:

- Total population
- Potential market-those in the total population who have interest in acquiring the product.
- Available market-those in the potential market who have enough money to buy the product.
- Qualified available market-those in the available market who legally are permitted to buy the product.
- Target market-the segment of the qualified available market that the firm has decided to serve (the *served market*).
- Penetrated market-those in the target market who have purchased the product.

The size of the market is not necessarily fixed. For example, the size of the available market for a product can be increased by decreasing the product's price, and the size of the qualified available market can be increased through changes in legislation that result in fewer restrictions on who can buy the product.

Defining the market is the first step in analysing it. Since the market is likely to be composed of consumers whose needs differ, market segmentation is useful in order to better understand those needs and to select the groups within the market that the firm will serve.

THE MARKETING PROCESS

Under the marketing concept, the firm must find a way to discover unfulfilled customer needs and bring to market products that satisfy those needs. The process of doing so can be modeled in a sequence of steps: the situation is analysed to identify opportunities, the strategy is formulated for a value proposition, tactical decisions are made, the plan is implemented and the results are monitored. A thorough analysis of the situation in which the firm finds itself serves as the basis for identifying opportunities to satisfy unfulfilled customer needs. In addition to identifying the customer needs, the firm must understand its own capabilities and the environment in which it is operating.

The situation analysis thus can be viewed in terms an analysis of the external environment and an internal analysis of the firm itself. The external environment can be described in terms of macro-environmental factors that broadly affect many firms, and micro-environmental factors closely related to the specific situation of the firm. The situation analysis should include past, present, and future aspects. It should include a history outlining how the situation evolved to its present state, and an analysis of trends in order to forecast where it is going. Good forecasting can reduce the chance of spending a year bringing a product to market only to find that the need no longer exists. If the situation analysis reveals gaps between what consumers want and what currently is offered to them, then there may be opportunities to introduce products to better satisfy those consumers. Hence, the situation analysis should yield a summary of problems and opportunities. From this summary, the firm can match its own capabilities with the opportunities in order to satisfy customer needs better than the competition. There are several frameworks that can be used to add structure to the situation analysis:

- 5 C Analysis-company, customers, competitors, collaborators, climate. Company represents the internal situation; the other four cover aspects of the external situation
- PEST analysis-for macro-environmental political, economic, societal, and technological factors. A PEST analysis can be used as the "climate" portion of the 5 C framework.

- SWOT analysis-strengths, weaknesses, opportunities, and threats-for the internal and external situation. A SWOT analysis can be used to condense the situation analysis into a listing of the most relevant problems and opportunities and to assess how well the firm is equipped to deal with them.

Once the best opportunity to satisfy unfulfilled customer needs is identified, a strategic plan for pursuing the opportunity can be developed. Market research will provide specific market information that will permit the firm to select the target market segment and optimally position the offering within that segment. The result is a value proposition to the target market. The marketing strategy then involves:

- Segmentation
- Targeting (target market selection)
- Positioning the product within the target market
- Value proposition to the target market.

Marketing Mix Decisions

Detailed tactical decisions then are made for the controllable parameters of the marketing mix. The action items include:

- Product development-specifying, designing, and producing the first units of the product.
- Pricing decisions
- Distribution contracts
- Promotional campaign development.

At this point in the process, the marketing plan has been developed and the product has been launched. Given that few environments are static, the results of the marketing effort should be monitored closely. As the market changes, the marketing mix can be adjusted to accomodate the changes.

Often, small changes in consumer wants can addressed by changing the advertising message.

As the changes become more significant, a product redesign or an entirely new product may be needed. The marketing process does not end with implementation-continual monitoring and adaptation is needed to fulfil customer needs consistently over the long-term.

SITUATION ANALYSIS

In order to profitably satisfy customer needs, the firm first must understand its external and internal situation, including the customer, the market environment, and the firm's own capabilities. Furthermore, it needs to forecast trends in the dynamic environment in which it operates. A useful framework for performing a situation analysis is the 5 C Analysis. The 5C analysis is an environmental scan on five key areas especially applicable to marketing decisions.

It covers the internal, the micro-environmental, and the macro-environmental situation. The 5 C analysis is an extension of the 3 C analysis (company, customers, and competitors), to which some marketers added the 4th C of collaborators.

The further addition of a macro-environmental analysis (climate) results in a 5 C analysis, some aspects of which are outlined below.

Company

- Product line
- Image in the market
- Technology and experience
- Culture
- Goals.

Collaborators

- Distributors
- Suppliers
- Alliances.

Customers

- Market size and growth
- Market segments
- Benefits that consumer is seeking, tangible and intangible.
- Motivation behind purchase; value drivers, benefits vs. costs
- Decision maker or decision-making unit
- Retail channel-where does the consumer actually purchase the product?
- Consumer information sources-where does the customer obtain information about the product?
- Buying process; e.g. impulse or careful comparison
- Frequency of purchase, seasonal factors
- Quantity purchased at a time
- Trends-how consumer needs and preferences change over time.

Competitors

- Actual or potential
- Direct or indirect
- Products
- Positioning
- Market shares
- Strengths and weaknesses of competitors.

Climate (or context)

The climate or macro-environmental factors are:

- Political & regulatory environment-governmental policies and regulations that affect the market
- Economic environment-business cycle, inflation rate, interest rates, and other macroeconomic issues
- Social/Cultural environment-society's trends and fashions
- Technological environment-new knowledge that makes possible new ways of satisfying needs; the impact of technology on the demand for existing products.

Information Sources

Customer and competitor information specifically oriented toward marketing decisions can be found in market research reports, which provide a market analysis for a particular industry. For foreign markets, country reports can be used as a general information source for the macro-environment. By combining the regional and market analysis with knowledge of the firm's own capabilities and partnerships, the firm can identify and select the more favourable opportunities to provide value to the customer.

4

Collection of Data and Organization

DATA COLLECTION

Data Collection helps your team to assess the health of your process. To do so, you must identify the key quality characteristics you will measure, how you will measure them, and what you will do with the data you collect.

What exactly is a key quality characteristic? It is a characteristic of the product or service produced by a process that customers have determined is important to them.

Key quality characteristics are such things as the speed of delivery of a service, the finish on a set of stainless steel shelves, the precision with which an electronic component is calibrated, or the effectiveness of an administrative response to a tasking by higher authority.

Every product or service has multiple key quality characteristics. When you are selecting processes to improve, you need to find out the processes, or process steps, that produce the characteristics your customers perceive as important to product quality.

Data Collection is nothing more than planning for and obtaining useful information on key quality characteristics produced by your process. However, simply collecting data does not ensure that you will obtain relevant or specific enough data to tell you what is occurring in your process. The key issue is not: How do we collect data? Rather, it is: How do we obtain useful data?

STATISTICS/METHODS OF DATA COLLECTION

The main portion of Statistics is the display of summarized data. Data is initially collected from a given source, whether they are experiments, surveys, or observation, and is presented in one of four methods:

Textular Method

The reader acquires information through reading the gathered data.

Tabular Method

Provides a more precise, systematic and orderly presentation of data in rows or columns.

Semi-tabular Method

Uses both textual and tabular methods.

Graphical Method

The utilization of graphs is most effective method of visually presenting statistical results or findings.

Experiments

It's true that scientists utilize two methods to gather information about the world: correlations (a.k.a. observational research) and experiments.

Experiments, no matter the scientific field, all have two distinct variables. Firstly, an independent variable (IV) is manipulated by an experimenter to exist in at least two levels (usually "none" and "some"). Then the experimenter measures the second variable, the dependent variable (DV).

A simple example—

Suppose the experimental hypothesis that concerns the scientist is that reading a Wiki will enhance knowledge. Notice that the hypothesis is really an attempt to state a causal relationship like, "if you read a Wiki, then you will have enhanced knowledge." The antecedent condition (reading a Wiki) causes the consequent condition (enhanced knowledge). Antecedent conditions are always IVs and consequent conditions are always DVs in experiments. So the experimenter would produce two levels of Wiki reading (none and some, for example) and record knowledge. If the subjects who got no Wiki exposure had less knowledge than those who were exposed to Wikis, it follows that the difference is caused by the IV.

So, the reason scientists utilize experiments is that it is the only way to determine causal relationships between variables. Experiments tend to be artificial because they try to make both groups identical with the single exception of the levels of the independent variable.

Sample surveys involve the selection and study of a sample of items from a population. A sample is just a set of members chosen from a population, but not the whole population. A survey of a whole population is called a census. A sample from a population may not give accurate results but it helps in decision making.

Examples

Examples of sample surveys:

- Phoning the fifth person on every page of the local phonebook and asking them how long they have lived in the area. (Systematic Sample).
- Dropping a quad. in five different places on a field and counting the number of wild flowers inside the quad. (Cluster Sample).
- Selecting sub-populations in proportion to their incidence in the overall population. For instance, a researcher may have reason to select a sample consisting 30% females and 70% males in a population with those same gender proportions. (Stratified Sample).
- Selecting several cities in a country, several neighbourhoods in those cities and several streets in those neighbourhoods to recruit participants for a survey (Multi-stage sample).

The term random sample is used for a sample in which every item in the population is equally likely to be selected.

Bias

While sampling is a more cost effective method of determining a result, small samples or samples that depend on a certain selection method will result in a bias within the results.

The following are common sources of bias:

- Sampling bias or statistical bias, where some individuals are more likely to be selected than others (such as if you give equal chance of cities being selected rather than weighting them by size).
- Systemic bias, where external influences try to affect the outcome (e.g. funding organizations wanting to have a specific result).

SURVEY DATA AND QUESTIONNAIRES

Survey data can be gathered in many ways, such as with the use of questionnaires administered in person, interviews, and telephone surveys. Information attained through surveys benefits researchers by recording personal behaviours, attitudes, identities, and definitions from a high volume of respondents. Surveys are useful for gaining information that is difficult to discuss or cannot otherwise be calculated by observation, such as risk, fear, religious belief, and racial attitude. Ulmer and Wilson (2003) believe that the strength of using surveys comes from the fact that "the researcher can rely on outside observations. As a result, many scholars believe that surveys, under certain circumstances, provide more valid measures of phenomena than do official records". Moreover, surveys provide a systematic and structured method for acquiring information on the same topic from a large group of people in a relatively short amount of time. However, survey data does have its limits. Surveys require that respondents answer questions honestly and accurately, and researchers must be aware that questions are subject to interpretation by the respondents. Therefore, it is recommended that researchers check with respondents on their answers to make sure there is as much uniformity in their interpretation as possible, employ surveys with a published record of reliability, and determine that questions asked in the survey accurately represent the study's intentions.

Esposito and Murphy (1999) believe the quantitative data typically acquired through survey responses do not give a very deep interpretation of occurrences, such as open-ended qualitative interviews would provide. These "snapshot", point-in-time measures give a shallow view that fails to give a rich or true assessment of the aspect of life being studied.

In order to study life for research purposes, access must be obtained to the information. The nature of the research and questions asked help establish the types of access required. Remenyi, Williams, Money, and Swartz (1998) believe that good research sites should have an entrance, trust can be established, and possibilities for gathering credible/quality and suitable evidence are available.

Finally, questions concerning data collection should analyse whether the strategy for collection has been clearly stated or not. Without a clearly articulated method for data collection, replications are obstructed and validity and reliability might come into question. These questions should address whether or not the best approach has been chosen for answering the primary research question.

Mixed Methods

Although mixed methods is not within the scope of the question addressed in the beginning of this chapter, it is being briefly described here, because of its apparent utility in survey research which is addressed as the focus of this chapter. Also, it has been included in this piece out of hindsight and necessity for possible usage in a proposed-future dissertation by the author.

Determining whether to use qualitative or quantitative research should become more apparent as the research process unfolds. Hines (2000) believes that multiple methods provide construct validity, as well as internal and external validity, while allowing complex issues to be examined using the respondents' language. Creswell (2003) also indicates that using mixed-methods research will assist in making better interpretations because informants will be providing information that is both measurable and analysed through rich description. A mixed-methods approach will also aid in reducing researcher bias and allowing documentation to be measured and analysed more effectively.

Likewise, Malterud (2001) states that, "qualitative studies can also be added to quantitative ones, to gain a better understanding of the meaning and implications of the findings".

Qualitative data is what often shapes the body of the research. Furthermore, it depicts the world as having several realities rather than one truth. On the other hand, quantitative research allows the use of data that identifies partial regularities, turnover in numbers, and statistical operations. Quantification shows a mixing and churning in numeric or sequential patterns and also emulates natural science. The practice of gathering and processing qualitative data should include the supplemental use of statistical hypothesis testing or comparing data sets with other samples and tests in order to confirm patterns.

Determining whether or not to use qualitative or quantitative research is not always so cut and dry. When great amounts of data are sought, quantitative research is often the most efficient and cost-effective research method. However, when data analysis and research questions require large amounts of description dealing with human experiences and perceptions, quantitative research could restrict the findings, making a qualitative approach ideal. The study itself should determine the method.

Ulmer and Wilson (2003) argue that "researchers can use qualitative data and statistical analysis without violating pragmatist ontological and epistemological positions" by meeting the following five conditions that serve to incorporate methods: (a) Using quantification and statistical analysis of outcomes of social processes to sensitize or lay the groundwork for qualitative study of the social processes in question themselves; (b) understanding that quantification should not be seen as synonymous with empiricism; (c) specifying that individual and joint actors, not variables, exercise agency (i.e., variables do not think or do things, people do, and we create variables to represent those things); (d) remembering that social causality lies not in variables or statistical models but in interpretive processes as people individually and jointly define situations and act within them; and (e) striving to keep the conceptual distance between the quantified measures and the phenomena measured as small as possible (i.e., maximizing validity).

As Hines (2000) asserts, there is no one best method for conducting survey research. Qualitative research is valuable for describing events, circumstances, and other areas of human understanding, including people's experiences and meanings in a normal social setting, as well as processes and structures. Similarly, Bryman (1993) believes qualitative research is best defined in studies of the social world; as it analyses and describes human behaviour from the point of view of those being studied. However, quantitative research too has value due to its ability to measure data, generalize findings, and make predictions.

These research strategies should not be thought of as incompatible, but complementary. Their procedures for interpreting data and other types of information are interpreted differently, and they answer different types of questions. Their underlying principle for increasing understanding of a subject, however, is the same, and they can be used to interpret or enhance one another. Mixed-methods research utilizes both qualitative and quantitative methods in order to explore a topic in breadth and depth. Mixed methods lead to a "truer analysis," as Davies (2003) indicates, and quantitative discoveries should lead to a reflection on qualitative decision making in order to get a better understanding of the ideas in the "heads of economic actors and the models they use".

Research should occur through multiple methods of gathering data for the purpose of drawing a robust conclusion. As Davies (2003) explains, incorporation of both qualitative and quantitative designs in research could yield results that would not have been identified through use of one method or the other. Mixed method clearly has their advantages in research and is often the answer to the debate over which type of research is best.

Although not every tool and method in either quantitative or qualitative research was discussed in this chapter, it is clear that each method represents a different paradigm, making no possible answer as to which is conclusively best. One primary conclusion that can be drawn from this chapter is that for any study proposed, one will not know which research method will produce the most accurate conclusions without examining the kind of knowledge sought. Once a survey is initiated and the data is collected, clues to new avenues that researchers should pursue and new questions to be asked will be revealed. During this process, researchers need to be aware of personal biases in collecting and analysing data. They should find ways to keep their own and their respondents' biases in check in order to produce results that accomplish what scholars have yearned to do for centuries—help society grow in knowledge through realized truths.

DATA PREPARATION

Data Preparation involves checking or logging the data in; checking the data for accuracy; entering the data into the computer; transforming the data; and developing and documenting a database structure that integrates the various measures.

Logging the Data

In any research project you may have data coming from a number of different sources at different times:

- mail surveys returns
- coded interview data
- pretest or posttest data
- observational data.

In all but the simplest of studies, you need to set up a procedure for logging the information and keeping track of it until you are ready to do a comprehensive data analysis. Different researchers differ in how they prefer to keep track of incoming data. In most cases, you will want to set up a database that enables you to assess at any time what data is already in and what is still outstanding. You could do this with any standard computerized database program (e.g., Microsoft Access, Claris Filemaker), although this requires familiarity with such programs. or, you can accomplish this using standard statistical programs (e.g., SPSS, SAS, Minitab, Datadesk) and running simple descriptive analyses to get reports on data status. It is also critical that the data analyst retain the original data records for a reasonable period of time — returned surveys, field notes, test protocols, and so on. Most professional researchers will retain such records for at least 5-7 years. For important or expensive studies, the original data might be stored in a data archive. The data analyst should always be able to trace a result from a data analysis back to the original forms on which the data was collected. A database for logging incoming data is a critical component in good research record-keeping.

Checking the Data For Accuracy

As soon as data is received you should screen it for accuracy. In some circumstances doing this right away will allow you to go back to the sample to clarify any problems or errors. There are several questions you should ask as part of this initial data screening:

- Are the responses legible/readable?
- Are all important questions answered?
- Are the responses complete?
- Is all relevant contextual information included (e.g., data, time, place, researcher)?

In most social research, quality of measurement is a major issue. Assuring that the data collection process does not contribute inaccuracies will help assure the overall quality of subsequent analyses.

Developing a Database Structure

The database structure is the manner in which you intend to store the data for the study so that it can be accessed in subsequent data analyses. You might use the same structure you used for logging in the data or, in large complex studies, you might have one structure for logging data and another for storing it. As mentioned above, there are generally two options for storing data on computer – database programs and statistical programs. Usually database programs are the more complex of the two to learn and operate, but they allow the analyst greater flexibility in manipulating the data. In every research project, you should generate a printed codebook that describes the data and indicates where and how it can be accessed. Minimally the codebook should include the following items for each variable:

- variable name
- variable description
- variable format (number, data, text)
- instrument/method of collection
- date collected
- respondent or group
- variable location (in database)
- notes.

The codebook is an indispensable tool for the analysis team. Together with the database, it should provide comprehensive documentation that enables other researchers who might subsequently want to analyse the data to do so without any additional information.

Entering the Data into the Computer

There are a wide variety of ways to enter the data into the computer for analysis. Probably the easiest is to just type the data in directly. In order to assure a high level of data accuracy, the analyst should use a procedure called double entry. In this procedure you enter the data once. Then, you use a special program that allows you to enter the data a second time and checks each second entry against the first. If there is a discrepancy, the program notifies the user and allows the user to determine the correct entry. This double entry procedure significantly reduces entry errors. However, these double entry programs are not widely available and require some training. An alternative is to enter the data once and set up a procedure for checking the data for accuracy. For instance, you might spot check records on a random basis. Once the data have been entered, you will use various programs to summarize the data that allow you to check that all the data are within acceptable limits and boundaries. For instance, such summaries will enable you to easily spot whether there are persons whose age is 601 or who have a 7 entered where you expect a 1-to-5 response.

Data Transformations

Once the data have been entered it is almost always necessary to transform the raw data into variables that are usable in the analyses. There are a wide variety of transformations that you might perform. Some of the more common are:

- missing values: Many analysis programs automatically treat blank values as missing. In others, you need to designate specific values to represent missing values. For instance, you might use a value of-99 to indicate that the item is missing. You need to check the specific program you are using to determine how to handle missing values.
- item reversals: On scales and surveys, we sometimes use reversal items to help reduce the possibility of a response set. When you analyse the data, you want all scores for scale items to be in the same direction where high scores mean the same thing and low scores mean the same thing. In these cases, you have to reverse the ratings for some of the scale items. For instance, let's say you had a five point response scale for a self esteem measure where 1 meant strongly disagree and 5 meant strongly agree. One item is "I generally feel good about myself." If the respondent strongly agrees with this item they will put a 5 and this value would be indicative of higher self esteem. Alternatively, consider an item like "Sometimes I feel like I'm not worth much as a person." Here, if a respondent strongly agrees by rating this a 5 it would indicate low self esteem. To compare these two items, we would reverse the scores of one of them (probably we'd reverse the latter item so that high values will always indicate higher self esteem). We want a transformation where if the original value was 1 it's changed to 5, 2 is changed to 4, 3 remains the same, 4 is changed to 2 and 5 is changed to 1. While you could program these changes as separate statements in most program, it's easier to do this with a simple formula like:

New Value = (High Value + 1)-Original Value.

In our example, the High Value for the scale is 5, so to get the new (transformed) scale value, we simply subtract each Original Value from 6 (i.e., 5 + 1).

- scale totals: Once you've transformed any individual scale items you will often want to add or average across individual items to get a total score for the scale.
- categories: For many variables you will want to collapse them into categories. For instance, you may want to collapse income estimates (in dollar amounts) into income ranges.

TECHNIQUES OF DATA COLLECTION

Basic requirements for scientific data are that it should be reliable and impartial. In Sociology these conditions are hard to meet. Yet numerous methods are used to minimize errors in data. Some of the commonly used sources in collecting data are:

- Existing materials including the official statistical record and historical and contemporary documents.
- Social surveys through questionnaire and schedules
- Interviewing
- Observation-Participants and non-participant.

Existing Material

Statistical Sources: Government statistics particularly census or statistics produced by large industrial or commercial firms, trade unions or other organizations provide one important account of data which sociologist can use in their analysis. An outstanding example of the imaginative use of official statistics in the positivist tradition is the study of suicide made by the famous French sociologist Emile Durkheim in the 19th century. However official statistics are the kind of data that are not collected by sociologists themselves and so there problems while analysing the data.

Historical documents: Records and accounts of qualitative kind for example relating to belief, values, social relationship or social behaviour may also be contemporary or may refer to earlier periods.

There are several difficulties immediately present themselves in the use of records from the past. Few chroniclers of social relation and social action record observations in the systematic way in which the sociologists are interested. There are often intriguing and sympathetic records but the information that is vital to the sociologist is often missing.

Contemporary Records: Contemporary records relating to social relationship and social behaviour are seldom used as the sole source of information and sociological research. They are usually one source of a particular account or achievement.

Social Survey

The basic procedure in survey is that people are asked a number of questions on that aspect of behaviour which the sociologist is interested in. A number of people carefully selected so that their representation of their population being studied are asked to answer exactly the same question so that the replies to different categories of respondents may be examined for differences. One type of survey relies on contacting the respondents by letter and asking them to complete the questionnaire themselves before returning it. These are called Mail questionnaires. Sometimes questionnaires are not completed by individuals separately but by people in a group under the direct supervision of the research worker. A variation of the procedure can be that a trained interviewer asks the questions and records the responses on a schedule from each respondent.

These alternate procedures have different advantages and disadvantages. Mail questionnaires are relatively cheap and can be used to contact respondents who are scattered over a wide area. But at the same time the proportion of people who return questionnaires sent through post is usually rather small. The questions asked in main questionnaires have also to be very carefully worded in order to avoid ambiguity since the respondents cannot ask to have questions clarified for them. Using groups to complete questionnaires means that the return rate is good and that information is assembled quickly and fairly. Administrating the interview schedules to the respondents individually is probably the most reliable method. Several trained interviewers may be employed to contact specific individuals. The questionnaires and schedules can consist of both close-ended and open-ended questions. Also a special attention needs to be paid to ensure that the questionnaires are filled in logical order.

Where aptitude questions are included great care must be exercised to ensure the proper words are used. In case of schedules emphasis and interactions may also be standardized between different individuals and from respondents to respondents. Finally proper sampling techniques must be used to ensure that the sample under study represents the universe of study. In order to enhance the reliability of data collected through questionnaires and schedules, these questionnaires and schedules must be pretested through pilot studies.

Interviewing

Social surveys may depend either on questionnaires that are self-administered or on schedules completed by trained research workers personally interviewing then is not a method of data collection distinct from social surveying but rather a technique which may vary from the brief formal contact as when the interviewer is working for the firms public opinion consultants or a market research organization and simply asks a housewife a few highly specific questions on limited range of topics to a long interview in which the research worker allows the respondents to develop points at leisure and take up others as he chooses.

The brief formal interview in which the working of the questions and the order in which they are asked is fixed is called structured interview while the freer discursive interview is called unstructured interview. The object of using structured interview is to standardize the interview as much as possible and thus to reduce the effect that the interviewer's personal approach or biases may have upon the result and even when structured interviews are used, proper training can do a lot to ensure further the reliability and validity of research. The personality of the interviewer and the social characteristics that the respondents attribute to him can be having influence on the result.

The effort of interviewer's bias can be estimated by comparing one interviewer's result with other. The problem of interviewer's bias in an unstructured interview is much greater. Here the interviewer is left to his common devices as far as the way he approaches a respondent is concerned. There is no fixed list of questions to work through. Instead the interviewer may work from a guide that will remind him of the topics he wishes to cover. The training of the interviewer is crucial here not simply training in the social skills of keeping the conversation going on a topic that the respondent may not be very interested in but also in acquiring sensitivity to those things his respondents tells him which are specially relevant to the theoretical topics he is pursuing. This means that unstructured interviews can be carried out by people trained in sociological theory. They are then able to size upon stray comments made by the respondents which can be developed and lead on to important theoretical insight.

Observation: Participant and non Participant

The rationale behind the use of observation in sociological research is that the sociologist should become party to a set of social actions sufficiently able to be able to assess directly the social relationship involved. The degree of involvement may vary considerable from being merely a watcher on the sidelines to be deeply involved in and being a part of what is going on. The former type of observation techniques are called non-participant while the latter is called participant observation. Sometimes one way observations screen have been used to watch groups in actions that they are unaware that they are being watched and the observer cannot affect their actions by his presence. The sociologist is visibly present and is a part of the situation either as a sociologist or in another guise. Where the sociologist is merely an observer it is usually assumed that he knows enough about what the actors are doing to be able to understand their behaviour. Any sociological observer has then to some extent be a participant observer he must at least share sufficient cultural background with the actors to be able to construe their behaviour meaningfully but the degree of participation and of sharing of meaning may vary considerably. Examples of such studies are Nel Anderson's study of Hobo-Indians and William White study of Street Corner Society.

Sampling

For practical and cost reasons, it is often impossible to collect information about the entire population of people or things in which social researchers are interested. In these cases, a sample of the total is selected for study. Most statistical studies are based on samples and not on complete enumerations of all the relevant data.

The main criteria when sampling are to ensure that a sample provides a faithful representation of the totality from which it is selected, and to know as precisely as possible the probability that a sample is reliable in this way. Randomization meets these criteria, because it protects against bias in the selection process and also provides a basis on which to apply statistical distribution theory that allows an estimate to be made of the probability that conclusions drawn from the sample are correct. A statistical sample is a miniature picture or cross-section of the entire group or aggregate from which the sample is taken. The entire group from which a sample is chosen is known as the population, universe or supply.

Simple Random Sampling

The basic type of random sample is known as a simple random sample, one in which each person or item has an equal chance of being chosen. Often a population contains various distinct groups or strata that differ on the attribute that is being researched.

Stratified Random Sampling

Stratified random sampling involves sampling of each stratum separately. This increases precision, or reduces time, effort and cost of allowing smaller sample sizes for a given level of precision. For example, poverty is known to be most common among the elderly, the unemployed and single parent families, so research on the effect of poverty might will sample separately each of these three strata as part of a survey of poverty in the population as a whole which would permit the total sample size to be reduced because the investigator would know that the groups most affected by poverty were guaranteed inclusion.

Cluster Sampling

Cluster sampling is sometimes used when the population naturally congregates into clusters. For example, managers are clustered in organizations, so a sample of managers could be obtained by taking a random sample of organizations and investigating the managers in each of these. Interviewing or observing managers on this basis would be cheaper and easier than using a simple random sample of managers scattered across all organizations in the country. This is usually less precise than a simple random sample of the same size, but in practice the reduction in cost per element more than compensated for the decrease in precision.

Multi-stage Sampling

Sampling may be done as one process or in stages, known as multi-stage sampling. Multi-stage designs are common when populations are widely dispersed. Thus a survey of business managers might proceed by selecting a sample of corporations as first stage units, perhaps choosing these corporations with a probability proportionate to their size, and then selecting a sample of managers within these corporations at the second stage. Alternatively, a sample of individual factories or office buildings within each corporation could be chosen as the second stage units, followed by sample of managers in each of these as a third stage. Stratification can also be used in the design, if for example occupational sub-groups are known to differ from each other, by selecting state such as personnel, production, and finance management and sampling within each of these. For sampling to be representative, one needs a complete and accurate list of the first stage units that make up the relevant population, a basic requirement that is not always easily met. This forms the sampling frame. Selection from the frame is best done by numbering the items and using a table of random numbers to identify which items form the sample, though a quasi-random method of simply taking every item from the list is often appropriate. The reliability of a sample taken from a population can be assessed by the spread of the sampling distribution, measured by the standard deviation of this distribution, called the standard error. As a general rule, the larger is the size of the sample the smaller the standard error.

Area Sampling

In sampling of this kind small areas are designated as sampling units and the households interviewed include all or a specified fraction of those found in a canvass of these designated small areas. The basic sampling units or segments chosen may be relatively large or relatively small depending on such factors as the type of area being studied, population distribution, the availability of suitable maps and other information and the nature and desired accuracy of the data being collected.

DATA COLLECTION METHODS AND INSTRUMENTS

Measurement of Disability

Collecting data about persons with disabilities and their lives is difficult. It poses unique problems that data collection developers need to address in the design phase of the collection process. Two issues need to be addressed at the outset: what kind of data collection instrument should be used, and what unit of measurement should be employed.

Collection Instruments

The main types of instruments for collecting data about persons with disabilities are:

- Population censuses
- Sample surveys (either general social surveys or specific health and disability surveys)
- Administrative collections and registries.

Each of these tools can be used to measure aspects of disability in a population and each has its strengths and weaknesses.

The Population Census certainly has the advantage of providing complete population coverage. Unfortunately, it is difficult to collect accurate information about disability in a census since time constraints make it unlikely that more than 4 to 6 general disability questions can be asked. Censuses, in some instances, also undercount children with disabilities and people with mild or moderate disabilities (in cases where the response categories are limited to 'Yes or No' options only). Still, for a crude measure of disability, and in the absence of other collection instruments, the census is useful.

Sample surveys are shorter surveys designed to be administered to a sub-population selected by some other instrument (often a census) that focus on specific issues. They are often put into the field to answer specific questions about a population. As such, they provide the opportunity to ask more detailed questions about disability. More detailed information is useful in itself, of course, but it also helps to reduce the number of false positive and negative responses, thereby offering a more accurate prevalence measure. A sample survey may be an independent survey focusing entirely on disability, or a disability module added to an existing survey.

Administrative collections and registers are composed of data that is collected as part of the normal operation of some service or programme. An example is the information found on a client intake form. These collections provide useful information on the characteristics of people accessing disability services as well as details about the services provided. They cannot give an accurate measure of disability prevalence since there is no guarantee of coverage and they tend to incorporate double counting. The quality of administrative register data is closely related to the quality of the administrative system, in particular, how well it has been maintained and how closely the concepts align with the disability concepts of interest.

Choice of Selection and Measurement Unit

The second preliminary issue that needs to be addressed, whatever data collection instrument is chosen, is how to select the unit for which disability is to be measured. If the selection unit is the individual, then the individual will also be the measurement unit; if the selection unit is a collection of people-invariably a household-then, a decision has to be made as to whether the measurement unit is the household itself (that is, all individuals in the household) or some individual in the household that meets specified criteria (age, gender, unemployed, and/or others). These decisions depend in part on what kinds of data are needed. Is data required for the number of persons with disabilities and their characteristics, or for an estimate on the number of households that include individuals with disabilities? Data about individuals is important, but estimates at the household level are also useful for detailed analysis of living arrangements and access to help and assistance.

There are also issues of practicality and efficiency. Having the household as the selection unit means that the survey or interview can be conducted in a setting in which information about other people can be efficiently collected, even if only one member of the household is given the full interview. Questions asked of a single household member may reveal another individual with a disability, thereby increasing the sample of person with disability, without adding to the number of households screened.

A note of caution applies, however, if the private household is chosen as the measuring unit. In such a case, thought needs to be given to the fact that this choice excludes residential care units, retirement homes, and other institutions such as prisons. Excluding people living in institutions underestimates the levels of disability for older people and for those with particular types of disability such as psychiatric disabilities. For a complete picture, if feasible, household surveys should be supplemented with institutional surveys of disability.

Population Censuses

Many countries have collected information about disability in their national population censuses. The United Nations' 2001 Guidelines and Principles for the Development of Disability Statistics suggests the inclusion of a 'general' disability question in censuses in order to get a rough idea of prevalence. In countries without a good household survey program, the census may be the only possible source for estimating disability prevalence and gaining an approximate estimate of types of disability in the country.

The amount of information on disability that can be collected in the population census is very limited, and is often confined to a single question. With only one question, false positive and false negative responses are more common and no complete measure of the number of persons with disabilities is possible, especially among children and the elderly. Still, census data should be readily used, where available, to develop more detailed follow-up surveys.

Many countries use both short and long forms of census questionnaires. The short form is for complete coverage of core topics, such as age, sex and location, and may also have a question on disability. A probability sample is then selected for the long form to be administered. The long form includes all questions in the short form plus supplementary questions for more detailed coverage of selected topics. Questions more detailed than a single general question on disability may be included in the long form.

Sample Surveys

Sample surveys are good methods of collecting data about persons with disabilities. They use sample selection procedures to identify a sub-population that is representative of the total population, unlike censuses that are designed to enumerate every household or individual in the country.

While sample surveys are considerably cheaper than censuses, the size of the sample affects the quality of the data, which is particularly subject to sampling error. This can frustrate analyses that rely on cross-tabulation.

A small sample size also means that sample surveys cannot provide detailed information for very small geographical areas. Where such data is needed, census or administrative data collections or sophisticated data modelling techniques are more useful options.

Determining sample sizes required to produce statistics with an acceptable level of sampling error is the job of experienced survey methodologists. Unfortunately, the technical details of sample designs are beyond the scope of this manual. The UN Guidelines and Principles for the Development of Disability Statistics has some helpful comments on sample size for disability surveys. (As a rough guide, surveys that have sufficient size to yield valid unemployment estimates for a given geographic area are usually large enough to produce disability estimates for the same area for as long as the prevalence rates are similar.)

As mentioned, a sample survey for disability can either be an independent or dedicated survey, or a module to other surveys, such as a national health survey, a general social survey or labour force survey. Where these general social surveys are cyclical, it is very useful to have a disability module or a small set of disability questions added on.

As dedicated disability surveys: Surveys specifically dedicated to collect disability data are good sources of information about prevalence rates, the causes and types of disability, underlying health conditions, severity and duration of disability, and the use of and need for assistive devices, changes in environment, policies and public awareness on disability.

Dedicated disability surveys maximise the amount of information that can be collected to meet users' needs. Many countries have not attempted to conduct such disability surveys because they are more costly than simply adding a few questions into a census or already existing sample surveys.

As a module to other sample surveys: In many countries, national household surveys-covering topics such as health, education, living conditions, family income and expenditure, labour force participation, nutrition, time use, crime, and so on-are conducted on a regular basis. Whether every five years, annually, quarterly or monthly, these surveys try to identify short and long-term social trends.

A cost-efficient and effective way of collecting disability data is to add disability questions-or a 'disability module'-to one or several of these cyclical surveys. This approach has been used in many countries. Most commonly, disability modules have been added to labour force or health surveys, and sometimes to living conditions surveys.

There are advantages to on-going sample surveys rather than ad hoc or one-off sample surveys for disability. On-going surveys can make maximum use of the resources initially expended, including the time and other resources used in preparing for the data collection, as well as the trained personnel and other resources dedicated to collecting, processing, and analysing the data on a regular basis.

On-going survey programs also offer opportunities to learn from previous experiences so that the quality and usefulness of the information produced can be improved. They allow for measurement of change over time in key indicators such as frequency of types of disability, severity of disability, quality of life, opportunities and participation of persons with disabilities, and rehabilitation needs. These data can be exceedingly valuable for policy development and evaluation.

ADMINISTRATIVE COLLECTIONS

Besides surveys and censuses, administrative data collection (general purpose administrative systems or administrative registries) intended to serve persons with disabilities can be an important method of gathering information about these people and their characteristics. In this method, any information collected is organized and becomes part of normal service administration procedure such as the information gathered using standard client intake forms for health, rehabilitation, or social work services.

Administrative data collections can take several forms, depending on the nature of the service, the format used, the type of information collected, and the method and frequency of collection. Some examples are population registers, vital registration systems, social security systems, registries of occupational injuries, rehabilitation programmes, assistive device services, and other services specifically designed to serve the needs of persons with disabilities.

Administrative registries are databases of records of people with particular characteristics and set up as part of the administration of services to these individuals. Registers are either established during a registration survey or other point in time or they can be on-going and regularly updated.

Administrative records and registers, moreover, often provide unique information about persons with disabilities.

The information is usually collected for reasons concerning the administration of the programme or service, but with care, can prove to be useful.

For example, data about children and adolescents in special education programmes is an indication of participation rates in education; general invalid pension registries are often organized by disease or impairment; and domestic care allowances help to track rates of children and adults with severe disabilities.

Often these data are collected annually, which provides a source for trend data on the prevalence of impairments or disabilities.

Ministries, government departments, advocacy groups, and service providers often maintain administrative records which they use to monitor and evaluate programs and services. Given confidentiality concerns, only aggregated data may be available, but this can still be useful.

It must be kept in mind, however, that data from registration systems cannot provide information about persons with disabilities who need a service or programme but do not receive it. Data about unmet need has to come from other sources.

Since the information collected in administrative data collections is limited to people receiving services, or otherwise known to the service agency (as in the case of people on a waiting list), these collections have limited coverage. Therefore administrative collections, despite all their benefits, are not good sources for estimating overall disability prevalence.

What are the benefits? As compared with censuses and sample surveys, administrative data collections have some advantages for disability data users:

Data from administrative collections are generally available on a regular or on-going basis, as they are part of regularly updated information systems linked to a service.

As administrative data are collected as part of the day-to-day running of a service, they use fewer resources than special purpose surveys.

If the government or other fund provider requires data collection as a condition of on-going funding to a service, the data collected by that service will tend to be more reliable, complete and of higher quality.

Information from administrative data collections is likely to be useful to a wide range of stakeholders such as service providers, higher-level bureaucrats, disability consumer groups, and researchers.

How can administrative data be used? Administrative data collections provide information on numbers and characteristics of service users, and the type, quantity and cost of services provided. Depending on the detail of the dataset and the complexity of the data collection format, a substantial amount of information about services, service users and service providers can be collected in this manner. Administrative data of this sort can then be adopted by service agencies, planners, consumer advocacy groups, and funding department for a variety of purposes:

- to provide an evidence base to support planning for future service delivery (e.g., by monitoring trends in client age or disability type);
- to indicate whether identified groups of people (e.g., particular ethnic groups) are accessing services as much as expected;
- to monitor the cost-effectiveness of service provision;
- to support a budget submission for increased funding.

Moreover, with only minimal analysis, these data can answer basic administrative questions such as:

- How many people were supported by the service in a given time period, and what were their characteristics and support needs?
- What type of support was provided and received?
- What quantity of support-e.g., in terms of staff hours-was provided and received?
- From whom was the support received (in terms of details of the service agency, such as size, staff profile, and hours)?
- What was the cost (total, per service type, per client) of providing these services?
- What were the outcomes for clients? (Examples of outcome questions include: Were clients satisfied with the services received? Were there increased levels of client participation in key life areas such as employment?)

5

Organizational Structure and Systems

ORGANIZATIONAL STRUCTURE

First, organizational charts detail an organization's structure. It may be functional, in which work units are divided based on what they do and named after those functions (e.g., research and development, marketing, sales, etc.). The structure may be divisional, based on product, customers served, or geographic location. Finally, an organizational chart may represent a matrix structure, in which work units are organized by both function and division.

Organizational Hierarchy and Centralization

In addition to outlining the type of organizational structure, the organizational chart also indicates the number of management levels, whether the organizational structure is tall or flat, and the span of control at each level. Tall organizations have many levels of middle management and small spans of control. Each manager supervises and directs few employees, and the chain of command has many managers. Conversely, a flat organization has fewer management levels and larger spans of control. Because managers supervise more employees, employees tend to have more autonomy and discretion in their jobs.

Organizational hierarchy and the number of management levels often indicates the degree of centralization within an organization. Centralized organizations are those in which most of the decision making occurs by a few people at the top of the hierarchy. This typically creates a top-down management structure, in which top-level managers strongly control the direction of the workplace through their decisions and supervision.

Conversely, an organization with a decentralized structure allows greater decision-making and authority at lower organizational levels. Highly decentralized companies may have units that operate nearly independently of one another. The degree of hierarchy on an organization's chart normally will help one to determine the degree of centralization or decentralization within its structure. Typically, the taller the organization, the more centralized it is; flatter organizations generally require more decentralization, because managers each have broader spans of control and cannot direct and closely supervise so many people. Additionally, as previously described, the organization's structure may indicate the degree of centralization. Functional structures tend to be more centralized than do divisional structures.

Chain of Command

The vertical and horizontal lines connecting the rectangles on an organizational chart indicate reporting relationships and chain of command. That is, they indicate which employees are directly responsible for the supervision of others and who has ultimate accountability for a group of employees.

TYPES OF ORGANIZATIONAL PLANNING STRUCTURES

Organizational planning involves designing an organization's structure to maximize efficiency. This includes dividing a company up into different units, departments, and teams. Prior to this division, managers must consider a company's goals and business obstacles as well as alternative company structures. Business goals may include, for example, increasing the flow of information, promoting teamwork, and reducing redundancy. Next, managers must consider the different organizational structures and select the one that holds the greatest potential for eliminating problems with the current corporate structure or for bringing about the desired structural environment. The selection may be made by taking different organizational models into consideration, as well. Managers and executives generally divide an organization into different units based on one of the following six criteria:

Function

A manufacturing firm typically includes functional units such as engineering, production, finance, sales, and personnel. These different functions are controlled by managers who head each function. In this organizational structure, each function primarily focuses on its core tasks (e.g., production or finance).

Product

Large, diversified companies may find it advantageous to divide their tasks based on product groups, such as foodstuffs, farm chemicals, and pharmaceuticals. This organizational structure has the benefit of enabling each business unit to produce desired results. However, it can lead to high administrative costs and redundancy.

Customers

Companies may be divided into different units based on target customers. For example, a book publisher may be organized by retail bookstores, mail-order book stores, online book stores, and school system tier, such as elementary, middle school, high school, and higher education.

Geographic Location

Many sales and service companies use geographic location as the basis for creating departments within the organization. This organizational structure calls for members of each group to concentrate on particular locations for which they are responsible.

Process

Some companies are organized by process. Process organization is common in manufacturing and clerical companies. For example, natural gas companies have different units for exploration, production, and distribution. This type of organizational structure enables each unit to have its own specialists.

Matrix

Matrix organizational structures include not only general functional units like production, sales, and finance, but also product or geographic units. Company executives frequently oversee the product units directly. The product units, in turn, collaborate with and coordinate the functional units. By adopting the matrix arrangement, companies attempt to reap the benefits of the functional and the product or geographic structures, while bypassing the inefficient and redundant aspects of the product structure. A company with a matrix organizational structure has functional units such as development, production, and sales matched in a matrix by product or geographic location.

Company executives and managers must strive to select the organizational structure that best suits their fields of business, that offers the optimal amount of control, specialization, and cooperation, and that facilitates key business activities while also taking into consideration concerns for efficiency and effectiveness.

ORGANIZATIONAL THEORY AND STRUCTURES

Project succeed or fail based on people. Most common project management form is of a project team working within an existing functional organization.

The project team draws member from various functional sections and uses them to work on the project until it is completed.

The Project within an Existing Organization

Project team working within an existing organization structure, once completed project team disbanded or absorbed back into main organization.

Function of many elements including:

- The size of the project to the organization
- The relative status and importance of the project
- The resources that are made available
- The strategic fit of the project with the overall strategic objectives of the organization.

The Functional Structure

In functional arrangement, power or status is defined by a vertical hierarchy through the OBS. Projects are likely to improve systems, procedures, methods or products, and they would tend to be internal rather than external projects. May be established for limited periods in order to address specific demands.

Functional structure common in large organizations typical examples.

- Central government
- Local government
- Police forces
- The armed forces
- Most large private companies.

Typically of large organizations have continuous rolling programmes of similar repetitive or semi-repetitive work. Benefits of functional structure:

- Offer clear reliable reporting system and responsibilities are clearly defined.
- Mirror traditional authority structures – section heads with lines of authority running down chain of command to individual operational units
- Simple and straightforward and is compatible with basic human instincts. Specialize in area rather than develop range of different advanced skills.
- Repetitive so people can use their experience from one aspect to the next aspect, Knowledge transfer
- Specialist knowledge shared around between the various functional members. Build up a library of specialist knowledge
- Direct relationships with various support services. Functional mangers

agree works programmes and timetables directly with support services such as IT

- Preferred by highly inflexible organizations like government departments, hospital services. Clearly defined roles and responsibilities clearly defined chain of command.

Disadvantages of function structure:

- Inflexible. Result functional units difficult to innovate and respond to change
- Functional output tend to be primary objective of organization. Project structure attempts to operate within the functional structure considered as secondary importance
- Cross-functional activities are discouraged. Functional people stick to own specialization, avoid being involved in cross-functional activities
- Functional people tend to see function as their future. Career path tied to function rather than projects
- Functional sections develop sub-groups. Tend to develop boundaries act as barriers to effective communication
- Cannot easily respond to fluctuations in workload
- Demand a greater degree of central support.

The main drawback with a pure functional structure is the development of operational or organizational islands. Organizational Island is a segment of the organization within the overall organizational structure, tend to act as a semi-independent sector. Control and communication tend to flow down through various functional divisions. Little communication and co-operation across the functional divisions, this is inefficient as cross-transfer of co-operation allow formation of horizontal units.

Pure Project Structure

Self contained section or unit within an otherwise purely functional structure. Typically used for projects that are difficult to plan accurately and where resource requirement and provision levels cannot be accurately established beforehand.

Project team disbanded upon completion of its project and individual project team members will return to the pool for use on other projects. Benefits of Pure project structure:

* System is flexible and responsive to change
* Innovation and evolution are not restricted
* Operational cost quickly adjusted
* Project manger in charge of any particular project is sole charge of project complete authority and control over project resources. No requirement for negotiating with functional manager or project sponsor

* Project staff clear reporting change
* Communication lines much shorter
* Informal communication lines shorter develop more quickly and efficiently
* Authority with the project
* Project team member do not have any functional loyalties
* Centralized support much simpler and the overall sizes of support function smaller
* Easier to incorporate external consultants.
* Possible to execute a series of related projects.

Projects point of view, pure project system appears to be the best supporting structure. Disadvantages of Pure Project structure:

* Several projects running concurrently duplication of effort
* Initial operating costs may be high considerable time before projects completed
* Centralize direction is needed, from of command hierarchy. Higher levels of authority may have difficulty interfacing
* Project mangers (by definition) tend to think ahead
* A sense of competition develop between various project teams
* Project team members tend to have underlying concern about long term commitment
* Project deadlines may create a culture 'cut corners'; maintain good performance records
* Difficult to compare the performance of individual projects
* Most staff in project structure has some form of original functional specialization prolonged absence from functional section lead to specialization becoming diluted over time.

Matrix Structure (Internal or Non Executive Project Management Structure)

Compromises between pure project and pure functional forms. It encourages horizontal communication and accountability. Popular structure for organizations undertaking projects. Combines benefits of the functional organization with those of the pure project organization, same time eliminating the disadvantages. Matrix structure is the pure project structure overlaid on the functional divisions.

Main characteristic of an internal project management structure are:

Functional boundaries: Run vertically through the system, defines the border or line of delineation between the engineering faculty and other faculties act as barrier to communication only gateway is through interfaces.

Power or status boundaries: Run horizontally through the system, represented as a triangle.

Organizational islands: Vertical (functional) and horizontal (power) boundaries.

Project sponsor: Internal system the project manger and the functional manger are both using individual employees as a shared resources. Risk direct competition between the project manger and functional manger over such resources, this conflict can be controlled through a project sponsor – moderator on any potential conflict.

The project management chair: Three different organizational interfaces:

Project manager – project sponsor (subordinate-boss).

Project manager – functional manger (peer-peer).

Project manger – project team (boss-subordinate).

Interfaces: Horizontal and vertical boundaries act as barrier to communication and co-operation. Interfaces are like gateways through barriers. Tool for control communication – Interface management system (IMS).

Interface management: Is management of the processes of communication and action across and within the various organizational interfaces. Good communication systems are set up so information flows rapidly and accurately.

The process of bidding: Resource allocation proposal calculate the approximate cost of labour, plant and material required for the project to be executed.

Time recording and cost-centre charging: Time spent on individual actives charged to the correct project or functional cost centre. Good points of an internal project management system include:

* Project operates as a self-contained unit and project manger has executive control over its operation and development
* Reasonable access to various functional units can access specialist input to the project
* Flexibility and adaptability. respond quickly to change
* Project stays in touch with operational objective of the functional units.
* Internal project management within a matrix structure offers the best of both the project and functional system to some extent
* Effectively spreads the risk between project and functional profitability
* Allows efficient balancing of functional and project resources provided the necessary control systems in place. Resources can be switched between one aspect and the other to absorb changes in demand
* Internal project management promotes innovation and evolution with organization while retaining the functional foundation

Weaknesses of internal project management:

* Balancing project and functional responsibilities, conflict can occur where relationship is not so good
* Team member often do not like having two bosses
* Matrix overlap means some responsibilities are effectively shared between the project and functional units
* Project management is complex specialist range of skills, reporting to a project sponsor only makes the job more complex
* Project tends to be depleted of resources toward the end of the implementation life cycle. Make completion very difficult
* Project team member have difficulty re-adjusting to functional units
* Project sponsor introduces new and additional level of authority and control.

Mixed and Hybrid Structure

This type of organization results in the project being spun off as a subsidiary company as it develops.

Summary comment on Internal Projects

Choosing the most appropriate structure for a particular project depends on many elements and these should be considered :

- The project, its objective's task, location and required resources
- The existing structure of the organization
- Previous experience of the type of project
- The client and the contract
- The project life span.

The Project External to the Existing Organization

External Project teams can be comprised as follows:

- All internal people, but be managed by an external project management consultant
- A mixture of internal people and external consultants, all of who are managed by the consultant project manger
- All external consultants, some or all of whom have been appointed by consultant project manager.

Surrogacy in this sense refers to the extent to which the client whishes to delegate control or authority to running the project.

External (Executive) Project Management Structure

External project mangers more applicable to smaller organizations with variable workload. Project manager is the only team lead full authority and control over all components of the project team. Including control and co-ordination of the design team. Number of distinctive characteristics:

Multidisciplinary and shared loyalty group characteristic: Form of fees for their external project management system tend to be strongly multidisciplinary. Susceptible to shared loyalty characteristics, working for his own practice or company not the same as the objectives of the client conflict of loyalties between individual parts of the system more susceptible to the problems of differentiation and sentience than internal project management teams.

Fee structures: Open competitive fee structure, negotiated fee are now generally accepted. Use fee bid package approach.

Look at a outline of project make a bid in form of a plan or proposal. Hourly or percentage basis. Typical fee schedule.

- Completion of pre-contract works
- Completion of post-contract work
- Completion of final account.

External contractual linkages: Wide range of contractual agreement systems, higher degree of risk of non-performance.

Take one of 3 primary forms.

Completion contracts – one-off contract specified goods, service agreed cost by specified date.

Term contracts – long term agreement agrees to supply good agreed rate to an agreed standard for a fixed term.

Service-level agreement – (SLAs) where level of service is set rather than the performance.

Priced and arranged in 2 primary ways.

Competitive contract – put out tender.

Negotiated contract - not involved in direct competition, client negotiates price and condition with contractor.

Contracts can take a number of different forms:

Fixed price contract - cost of project agreed and fixed in some way in advance.

Cost or cost-plus contract - fixed fees vs. price used insufficient design or product information cost plus - contract fee is % of overall cost on final account.

Reimbursement contract. Contractor perform the project at contractor expense and claim amount plus a fee every month.

Target Price contract – target price plus a fee.

Contract types can take number of different forms.

Standard forms of contract – Precise terms & conditions relate clearly to specific performance, aim establish absolute requirement form completion of the contract.

- Clauses carefully worded to cover all aspects of the contract layout established procedures for virtually all eventualities.
- Specific provisions within standard forms of contract, such as direct terms and conditions relating to loss and expense and extensions of time.
- Set out specific level of damages for non compliance with various events ie late completion.
- Breaches - Clearly state remedial action & right in the event of non compliance (damages clause).
- Professional services contract – implied terms interpret according to professional body, actual loss incurred.
- Supply contracts.
- Subcontract agreements.
- Proforma contracts.
- Damages not set out beforehand & based on actual loss incurred.
- Less protection to the client therefore more risk.
- Carry professional indemnity insurance in event of negligence.
- either state directly or imply a liability for the consultant to act within the professional standards that are set by the relevant professional body.
- Breaches / negligence or other act professional failed to act according to standards set out by the professional body.
- Most professional bodies produce a book of professional practice, which lays out the minimum standards and codes of conduct that members must observe.
- consultant is commissioned under a professional services contract that pertains to a given professional body, then any practice under the contract is deemed to be executed in accordance with that book of standards.
- typical contractual links in an external system.
  * Client to project manager and other design team members.
  * Client to main contractor.
  * Client to service authorities.
  * Client to nominated subcontractors and suppliers.
  * Client to local authority.

ORGANISATIONAL BEHAVIOUR AND BASICS

Organisational behaviour is an emerging filed of study, with sound prospects. Though the study of behaviour has interested psychologists for a number of years, its role in specific organizational situations has caught their attention only of late. The observation that work behaviour is instrumental in determining organizational performance has forced organizations to focus greater attention on their employees' behaviour. The trend is noticed to be greater than ever before. The traditional view of considering the employee as just a spoke in the Organizational wheel has almost virtually disappeared. Organizations are constantly infusing the much needed humanistic approaches to keep their employees spirits high. Organizational behaviour is significantly affected by several changes. Of prime importance, is the change in organizational structures. Bureaucratic models have given way to modern ones like the project and matrix designs. Analysing the stages in the evolution of organizational structures and their specific implications to behaviour in organizations are the focus of this unit. Every organization has certain objectives or goals which it strives to accomplish. In order to do so, people work together in groups. The activities of each of these individuals and groups are co-ordinated by authority-responsibility relationships.

These are determined on the basis of the hierarchy of the jobs. This is what we mean by "organization". Such clearly marked relationships help to ensure effective work performance. Some of the commonly used definitions of the term are given here, so as to draw a clearer picture of the term and its use. An organization is a group of individuals co-ordinated into different levels of authority and segments of specialization for the purpose of achieving the goals and objectives of the organization. The process of identifying and grouping the work to be performed, defining and delegating responsibility and authority, and establishing relationships for the purpose of enabling people to work most effectively together in establishing of objectives.

An organization is the formation of an effective machine, the management of an effective executive and the administration of an effective direction. Administration determines the goals, management strives towards it. Administrator determines the organization; the management uses it. An organization is a machine of management in its achievement of the ends determined by its administration. An analysis of these definitions show that organizations are born out of a necessity for a clear, well defined system or structure, which people require for execution of their work responsibilities. It helps them to relate to each other, co-ordinate their activities, and achieve the organizational goals or objectives. It minimizes confusion, maintains free from pressures nurtures suitable work behaviour, and maximizes effectiveness.

ORGANISATIONAL EFFECTIVENESS

People in the organization can tilt either way toward growth or toward stagnation. The design of the system in which people work can significantly influence which way they go and how far. The more they tilt toward stagnation, the higher the probability of developing organization illness.

And in general, the more they orient themselves toward growth, the better for organizational health. Behavioural research scientists are of the opinion that organization can be made more humane and at the same time more effective.

It is possible to design organizational systems where individual participants can give of themselves and not given themselves up. In spite of the fact that some basic conflicts between the individual and the organization, there can be new organizational structures, heightened competence in interpersonal relation among the managers, and increased group effectiveness. On the other hand incompetent organizational structures create executive environments lacking in trust, openness, dynamism and results in ineffective problem solving, poor decision making and organizational illness.

Organisational effectiveness is the central theme of organization theory. No management can think of a theory of organization that does not include the concept of effectiveness. Everyone-academicians and corporate people-unanimously agree to the significance of effectiveness in organization. Organisational effectiveness represents a desirable attribute in organization.

Effectiveness could be defined as the degree to which organization realizes its goals. Organisational effectiveness is the extent to which an organization, given certain resources and means, achieves its objectives without placing undue strain on its members. Effectiveness is the ability of an organization to mobilize its centres of power for action-producing and adaptation.

Effectiveness of an organization can be seen in terms of survival of the organization. According to silver and Sherman, Organisational effectiveness is the extent to which defined and finite resources can achieve its growth and profit without destroying its internal resources. It is how a company meets its minimization of stoppage between the input of resources and the optimized output. He further supports his viewpoint by an analogy that, there are parts of a ship which taken by themselves sink. The engine would plunge into the bottom of the sea and so would the propeller. The steel of the hull would sink. But when the parts of the ship are put together, they float. Thus organizational effectiveness is getting it together, to keep the enterprise afloat and competitive.

According to Houck levis Organisational effectiveness is the tool to which a goal that has been set for an operating task is attained in terms of quality and completion. Warren. G. Bennils opines Organisational effectiveness is the system's capacity to survive, adapt, maintain and grow, regardless of the particular function it performs.

Analytics

Analysing the stages in the evolution of organizational structures and their specific implications to behaviour in organizations are the focus of this unit. Every organization has certain objectives or goals which it strives to accomplish. In order to do so, people work together in groups. The activities of each of these individuals and groups are co-ordinated by authority-responsibility relationships. These are determined on the basis of the hierarchy of the jobs. This is what we mean by "organization". Such clearly marked relationships help to ensure effective work performance. Some of the commonly used definitions of the term are given here, so as to draw a clearer picture of the term and its use. An organization is a group of individuals co-ordinated into different levels of authority and segments of specialization for the purpose of achieving the goals and objectives of the organization. The process of identifying and grouping the work to be performed, defining and delegating responsibility and authority, and establishing relationships for the purpose of enabling people to work most effectively together in establishing of objectives.

An organization is the formation of an effective machine, the management of an effective executive and the administration of an effective direction. Administration determines the goals, management strives towards it. Administrator determines the organization; the management uses it. An organization is a machine of management in its achievement of the ends determined by its administration. An analysis of these definitions show that organizations are born out of a necessity for a clear, well defined system or structure, which people require for execution of their work responsibilities. It helps them to relate to each other, co-ordinate their activities, and achieve the organizational goals or objectives. It minimizes confusion, nurtures suitable work behaviour, and maximizes effectiveness.

Organizational Behaviour Basics

Changes are taking place all around in organization. It is but natural that everything changes over time. What remains constant is change itself. Manager finds that changes need to be initiated at different levels within the organization –individual, group, structure, processes and subsystems coping with internal and external environmental changes.

By scanning the environment of the organization and deciphering how changes in the environment are likely to widen the gap between desired and actual state of affairs of organization such as productivity, customer and employer satisfactions, profits, the manager can be a change – agent for introducing planned changes in the organization. The term work change in the organization refers to any alteration that occurs in the work environment. Planned changes mean those changes which are effected in a planned manner after assessing the need for change and working out the details as to when and how they will be carried out. Planned change is also called proactive change. In contrast, reactive change is one which takes place in random fashion as a crisis situation develops.

For proactive or planned change to be initiated, manager shall be sensitive to the environmental changes affecting the organization by being boundary scanners so that organizational crisis situations can be averted. Planned changes or proactive change is purposeful or goal-directed. There are two widely recognized goals of the planned change. One, it seeks to improve the organizational ability to cope with, or adapt to, changing environment. Second, it seeks to change employee knowledge, attitude and behaviour. A few generalizations about change reflect its nature. Change in any part of the organization tends to affect the whole organization. Change is a human as well as a technical problem. Social equilibrium exists in the organization employees are adjusted. When change occurs in the organization, it requires employees to make new adjustments as the organization seeks new equilibrium. Regarding it is the objective of management to restore and maintain equilibrium of work group and personal adjustment which change upsets. Though manager is a primary change agent, employees, outside consultants may also be change agents.

TRADITIONAL ORGANIZATIONAL STRUCTURE

The emergence of the traditional organizational structure, this section provides additional detail regarding how this affected the practice of management. The structure of every organization is unique in some respects, but all organizational structures develop or are consciously designed to enable the organization to accomplish its work. Typically, the structure of an organization evolves as the organization grows and changes over time. Researchers generally identify four basic decisions that managers have to make as they develop an organizational structure, although they may not be explicitly aware of these decisions. First, the organization's work must be divided into specific jobs. This is referred to as the division of labour. Second, unless the organization is very small, the jobs must be grouped in some way, which is called departmentalization. Third, the number of people and jobs that are to be grouped together must be decided. This is related to the number of people that are to be managed by one person, or the span of control—the number of employees reporting to a single manager. Fourth, the way decision-making authority is to be distributed must be determined.

In making each of these design decisions, a range of choices are possible. At one end of the spectrum, jobs are highly specialized with employees performing a narrow range of activities, while at the other end of the spectrum employees perform a variety of tasks. In traditional bureaucratic structures, there is a tendency to increase task specialization as the organization grows larger. In grouping jobs into departments, the manager must decide the basis on which to group them. The most common basis, at least until the last few decades, was by function. For example, all accounting jobs in the organization can be grouped into an accounting department, all engineers can be grouped into an engineering department, and so on. The size of the groupings also can range from small to large depending on the number of people the managers supervise. The degree to which authority is distributed throughout the organization can vary as well, but traditionally structured organizations typically vest final decision-making authority by those highest in the vertically structured hierarchy. Even as pressures to include employees in decision-making increased during the 1950s and 1960s, final decisions usually were made by top management. The traditional model of organizational structure is thus characterized by high job specialization, functional departments, narrow spans of control, and centralized authority. Such a structure has been referred to as traditional, classical, bureaucratic, formal, mechanistic, or command and control. A structure formed by choices at the opposite end of the spectrum for each design decision is called unstructured, informal, or organic. The traditional model of organizational structure is easily represented in a graphical form by an organizational chart. It is a hierarchical or pyramidal structure with a president or other executive at the top, a small number of vice presidents or senior managers under the president, and several layers of management below this, with the majority of employees at the bottom of the pyramid. The number of management layers depends largely on the size of the organization. The jobs in the traditional organizational structure usually are grouped by function into departments such as accounting, sales, human resources, and so.

Basis for Departmentalization

Many organizations group jobs in various ways in different parts of the organization, but the basis that is used at the highest level plays a fundamental role in shaping the organization. There are four commonly used bases.

Functional Departmentalization

Every organization of a given type must perform certain jobs in order do its work. For example, key functions of a manufacturing company include production, purchasing, marketing, accounting, and personnel. The functions of a hospital include surgery, psychiatry, nursing, housekeeping, and billing. Using such functions as the basis for structuring the organization may, in some instances, have the advantage of efficiency. Grouping jobs that require the same knowledge, skills, and resources allows them to be done efficiently and promotes the development of greater expertise. A disadvantage of functional groupings is that people with the same skills and knowledge may develop a narrow departmental focus and have difficulty appreciating any other view of what is important to the organization; in this case, organizational goals may be sacrificed in favour of departmental goals.

In addition, coordination of work across functional boundaries can become a difficult management challenge, especially as the organization grows in size and spreads to multiple geographical locations.

Product Departmentalization

Large, diversified companies are often organized according to product. All the activities necessary to produce and market a product or group of similar products are grouped together. In such an arrangement, the top manager of the product group typically has considerable autonomy over the operation. The advantage of this type of structure is that the personnel in the group can focus on the particular needs of their product line and become experts in its development, production, and distribution. A disadvantage, at least in terms of larger organizations, is the duplication of resources. Each product group requires most of the functional areas such as finance, marketing, production, and other functions. The top leadership of the organization must decide how much redundancy it can afford.

Customer/Market Departmentalization

An organization may find it advantageous to organize according to the types of customers it serves. For example, a distribution company that sells to consumers, government clients, large businesses, and small businesses may decide to base its primary divisions on these different markets. Its personnel can then become proficient in meeting the needs of these different customers. In the same way, an organization that provides services such as accounting or consulting may group its personnel according to these types of customers.

6

Sampling in Research Methodology

SAMPLING (STATISTICS)

Sampling is that part of statistical practice concerned with the selection of a subset of individual observations within a population of individuals intended to yield some knowledge about the population of concern, especially for the purposes of making predictions based on statistical inference. Sampling is an important aspect of data collection.

Researchers rarely survey the entire population for two reasons : the cost is too high, and the population is dynamic in that the individuals making up the population may change over time.

The three main advantages of sampling are that the cost is lower, data collection is faster, and since the data set is smaller it is possible to ensure homogeneity and to improve the accuracy and quality of the data.

Each observation measures one or more properties (such as weight, location, colour) of observable bodies distinguished as independent objects or individuals. In survey sampling, survey weights can be applied to the data to adjust for the sample design. Results from probability theory and statistical theory are employed to guide practice. In business and medical research, sampling is widely used for gathering information about a population.

Process

The sampling process comprises several stages:

- Defining the population of concern
- Specifying a sampling frame, a set of items or events possible to measure
- Specifying a sampling method for selecting items or events from the frame

- Determining the sample size
- Implementing the sampling plan
- Sampling and data collecting.

Population Definition

Successful statistical practice is based on focused problem definition. In sampling, this includes defining the population from which our sample is drawn. A population can be defined as including all people or items with the characteristic one wishes to understand. Because there is very rarely enough time or money to gather information from everyone or everything in a population, the goal becomes finding a representative sample (or subset) of that population.

Sometimes that which defines a population is obvious. For example, a manufacturer needs to decide whether a batch of material from production is of high enough quality to be released to the customer, or should be sentenced for scrap or rework due to poor quality. In this case, the batch is the population.

Although the population of interest often consists of physical objects, sometimes we need to sample over time, space, or some combination of these dimensions. For instance, an investigation of supermarket staffing could examine checkout line length at various times, or a study on endangered penguins might aim to understand their usage of various hunting grounds over time. For the time dimension, the focus may be on periods or discrete occasions.

In other cases, our 'population' may be even less tangible. For example, Joseph Jagger studied the behaviour of roulette wheels at a casino in Monte Carlo, and used this to identify a biased wheel. In this case, the 'population' Jagger wanted to investigate was the overall behaviour of the wheel (i.e. the probability distribution of its results over infinitely many trials), while his 'sample' was formed from observed results from that wheel. Similar considerations arise when taking repeated measurements of some physical characteristic such as the electrical conductivity of copper.

This situation often arises when we seek knowledge about the cause system of which the *observed* population is an outcome. In such cases, sampling theory may treat the observed population as a sample from a larger 'superpopulation'. For example, a researcher might study the success rate of a new 'quit smoking' program on a test group of 100 patients, in order to predict the effects of the program if it were made available nationwide. Here the superpopulation is "everybody in the country, given access to this treatment"-a group which does not yet exist, since the program isn't yet available to all.

Note also that the population from which the sample is drawn may not be the same as the population about which we actually want information. Often there is large but not complete overlap between these two groups due to frame issues etc. Sometimes they may be entirely separate-for instance, we might study rats in order to get a better understanding of human health, or we might study records from people born in 2008 in order to make predictions about people born in 2009.

Time spent in making the sampled population and population of concern precise is often well spent, because it raises many issues, ambiguities and questions that would otherwise have been overlooked at this stage.

Sampling Frame

In the most straightforward case, such as the sentencing of a batch of material from production (acceptance sampling by lots), it is possible to identify and measure every single item in the population and to include any one of them in our sample. However, in the more general case this is not possible. There is no way to identify all rats in the set of all rats. Where voting is not compulsory, there is no way to identify which people will actually vote at a forthcoming election (in advance of the election). These imprecise populations are not amenable to sampling in any of the ways below and to which we could apply statistical theory. As a remedy, we seek a *sampling frame* which has the property that we can identify every single element and include any in our sample. The most straightforward type of frame is a list of elements of the population (preferably the entire population) with appropriate contact information. For example, in an opinion poll, possible sampling frames include:

- Electoral register
- Telephone directory.

Not all frames explicitly list population elements. For example, a street map can be used as a frame for a door-to-door survey; although it doesn't show individual houses, we can select streets from the map and then visit all houses on those streets.

The sampling frame must be representative of the population and this is a question outside the scope of statistical theory demanding the judgment of experts in the particular subject matter being studied. All the above frames omit some people who will vote at the next election and contain some people who will not; some frames will contain multiple records for the same person. People not in the frame have no prospect of being sampled. Statistical theory tells us about the uncertainties in extrapolating from a sample to the frame. In extrapolating from frame to population, its role is motivational and suggestive.

To the scientist, however, representative sampling is the only justified procedure for choosing individual objects for use as the basis of generalization, and is therefore usually the only acceptable basis for ascertaining truth. — Andrew A. Marino

It is important to understand this difference to steer clear of confusing prescriptions found in many web pages.

In defining the frame, practical, economic, ethical, and technical issues need to be addressed. The need to obtain timely results may prevent extending the frame far into the future.

The difficulties can be extreme when the population and frame are disjoint. This is a particular problem in forecasting where inferences about the future are made from historical data. In fact, in 1703, when Jacob Bernoulli proposed to Gottfried Leibniz the possibility of using historical mortality data to predict the probability of early death of a living man, Gottfried Leibniz recognized the problem in replying:

Nature has established patterns originating in the return of events but only for the most part. New illnesses flood the human race, so that no matter how many experiments you have done on corpses, you have not thereby imposed a limit on the nature of events so that in the future they could not vary. —Gottfried Leibniz

Kish posited four basic problems of sampling frames:

1. Missing elements: Some members of the population are not included in the frame.
2. Foreign elements: The non-members of the population are included in the frame.
3. Duplicate entries: A member of the population is surveyed more than once.
4. Groups or clusters: The frame lists clusters instead of individuals.

A frame may also provide additional 'auxiliary information' about its elements; when this information is related to variables or groups of interest, it may be used to improve survey design. For instance, an electoral register might include name and sex; this information can be used to ensure that a sample taken from that frame covers all demographic categories of interest. (Sometimes the auxiliary information is less explicit; for instance, a telephone number may provide some information about location.)

Having established the frame, there are a number of ways for organizing it to improve efficiency and effectiveness. It's at this stage that the researcher should decide whether the sample is in fact to be the whole population and would therefore be a census.

COEFFICIENT OF DISPERSION

For any data, it is always desirable that the measure of dispersion is less. A small value for the measure means that the values in the data are more or less consistent, centering on their average. Suppose it is required to compare the dispersion between two or more data. For this purpose any measure of dispersion cannot be used as such for two reasons. the two data under study may certain to different variables such as heights and weights of the individual. Number of marriages and ages of persons who have committed suicides etc and even if the variables in the two data are the same, their averages may be different. Hence one is in need of a relative measure of dispersion which will be free from the two differences cited above. Such a measure is the coefficient of dispersion which is defined as the ratio between measure dispersion and an average. This coefficient is a constant, rendering ready comparison of dispersion between data. There are four coefficients which are commonly referred, namely

1. Quartile coefficient of dispersion = Quartile Deviation/Mean
2. Coefficient of variation (C. V) = Standard Deviation/Mean x 100
3. Mean Deviation/Mean
4. Mean Deviation/Median of which the first two are the most frequently used coefficients.

ASSOCIATION OF ATTRIBUTES

In social sciences, we come across certain phenomena which are incapable of quantitative measurement. Blindness, deafness, religion; juvenile delinquency, marital status etc are some phenomena which are not measurable. Such characteristics are called attributes. In these cases, one can make only counting of individuals who possess or do not possess these attributes. In other words what can do is to state so many individuals are blind or so many non-blind. While dealing with one attribute the classification of data is done on the basis of presence or absence of the attribute. It is also absolutely essential that a clear-cut definition of the attribute under study is made because only such a definition paves way for the counting of the individuals possessing or not possessing the attribute. Two attributes are said to be associated only if they appear together in a great number of cades than is to be expected if they are independent. On the other hand, if the number of observed cases is less than the expected, under assumption of independence, attributes are associated. In order to ascertain whether the attributes are associated or not the following methods can be used.

1. Comparison of observed and expected frequencies.
2. Proportion method

3. Yule's coefficient of Association
4. Coefficient of colligation
5. Coefficient of contingency.

Regression

The dictionary meaning of the term 'regression' is the act of returning or going back. The term was first used by Sir Francis Galton in 1877 when he analysed the relationship between the height of fathers and sons. His study of the heights of fathers and sons revealed a very interesting relationship-tall fathers tend to have tall sons and short fathers, short sons, but the average height of the son of a group of tall fathers is less than that of the fathers and the average height of the sons of short fathers is greater than that of the fathers. The line describing this tendency to regress or step back was called by Galton a Regression Line. The term continues to be in use to represent the trend present but no longer has it necessarily carried the original implication of stepping back that Galton intended.

The statistical device that enables the estimation or prediction of the unknown values of one variable based on the known values of another variable is termed regression. Regression is the measure of average relationship between variables.

Linear Regression

In case there exists relationship between two variables x and y the plotted points in the scatter diagram lie concentrated around a curve and the relationship is said to be expressed by means of curvilinear regression. In the particular case, when the curve is a straight line, is called the line of regression and the regression is said to be linear. Once it is known that the relationship is expressible by means of straight line, a technique known as the Principles of Least Squares is adopted to fit the line on the basis of the given bivariate data.

A line of regression gives the most probable values of one variable for the given value of the other variable. For two series x and y there are two regression lines, one on assuming x as dependent and y dependent variable and the two on assuming x as dependent and independent variable.

STATISTICAL TERMS IN SAMPLING

Let's begin by defining some very simple terms that are relevant here. First, let's look at the results of our sampling efforts. When we sample, the units that we sample—usually people—supply us with one or more responses. In this sense, a response is a specific measurement value that a sampling unit supplies. The person is responding to a survey instrument and gives a response of '4'. When we look across the responses that we get for our entire sample, we use a statistic.

There are a wide variety of statistics we can use—mean, median, mode, and so on. In this example, we see that the mean or average for the sample is 3.75. But the reason we sample is so that we might get an estimate for the population we sampled from. If we could, we would much prefer to measure the entire population. If you measure the entire population and calculate a value like a mean or average, we don't refer to this as a statistic, we call it a parameter of the population.

The Sampling Distribution

So how do we get from our sample statistic to an estimate of the population parameter? A crucial midway concept you need to understand is the sampling distribution. In order to understand it, you have to be able and willing to do a thought experiment. Imagine that instead of just taking a single sample like we do in a typical study, you took three independent samples of the same population.

And furthermore, imagine that for each of your three samples, you collected a single response and computed a single statistic, say, the mean of the response. Even though all three samples came from the same population, you wouldn't expect to get the exact same statistic from each. They would differ slightly just due to the random "luck of the draw" or to the natural fluctuations or vagaries of drawing a sample. But you would expect that all three samples would yield a similar statistical estimate because they were drawn from the same population. Now, for the leap of imagination!

Imagine that you did an *infinite* number of samples from the same population and computed the average for each one. If you plotted them on a histogram or bar graph you should find that most of them converge on the same central value and that you get fewer and fewer samples that have averages farther away up or down from that central value. In other words, the bar graph would be well described by the *bell curve* shape that is an indication of a "normal" distribution in statistics.

The distribution of an infinite number of samples of the same size as the sample in your study is known as the sampling distribution. We don't ever actually construct a sampling distribution. Why not? You're not paying attention! Because to construct it we would have to take an *infinite* number of samples and at least the last time I checked, on this planet infinite is not a number we know how to reach. So why do we even talk about a sampling distribution? Now that's a good question!

Because we need to realise that our sample is just one of a potentially infinite number of samples that we could have taken. When we keep the sampling distribution in mind, we realise that while the statistic we got from our sample is probably near the center of the sampling distribution (because most of the samples would be there) we could have gotten one of the extreme samples just by the luck of the draw.

If we take the average of the sampling distribution—the average of the averages of an infinite number of samples—we would be much closer to the true population average—the parameter of interest. So the average of the sampling distribution is essentially equivalent to the parameter. But what is the standard deviation of the sampling distribution (OK, never had statistics? There are any number of places on the web where you can learn about them or even just brush up if you've gotten rusty. This isn't one of them. I'm going to assume that you at least know what a standard deviation is, or that you're capable of finding out relatively quickly). The standard deviation of the sampling distribution tells us something about how different samples would be distributed. In statistics it is referred to as the standard error (so we can keep it separate in our minds from standard deviations. Getting confused? Go get a cup of coffee and come back in ten minutes...OK, let's try once more... A standard deviation is the spread of the scores around the average in a single sample. The standard error is the spread of the averages around the average of averages in a sampling distribution. Got it?)

Sampling Error

In sampling contexts, the standard error is called sampling error. Sampling error gives us some idea of the precision of our statistical estimate. A low sampling error means that we had relatively less variability or range in the sampling distribution. But here we go again—we never actually see the sampling distribution! So how do we calculate sampling error? We base our calculation *on the standard deviation of our sample*. The greater the sample standard deviation, the greater the standard error (and the sampling error).

The standard error is also related to the sample size. The greater your sample size, the smaller the standard error. Why? Because the greater the sample size, the closer your sample is to the actual population itself. If you take a sample that consists of the entire population you actually have no sampling error because you don't have a sample, you have the entire population. In that case, the mean you estimate is the parameter.

The 68, 95, 99 Per cent Rule

You've probably heard this one before, but it's so important that it's always worth repeating... There is a general rule that applies whenever we have a normal or bell-shaped distribution. Start with the average—the center of the distribution. If you go up and down (*i.e*, left and right) one standard unit, you will include approximately 68% of the cases in the distribution (*i.e*, 68% of the area under the curve). If you go up and down two standard units, you will include approximately 95% of the cases. And if you go plus-and-minus three standard units, you will include about 99% of the cases. Notice that I didn't specify in the previous few sentences whether I was talking about standard deviation units or standard error units.

That's because the same rule holds for both types of distributions (*i.e*, the raw data and sampling distributions). For instance, the mean of the distribution is 3.75 and the standard unit is 25 (If this was a distribution of raw data, we would be talking in standard deviation units. If it's a sampling distribution, we'd be talking in standard error units). If we go up and down one standard unit from the mean, we would be going up and down.25 from the mean of 3.75. Within this range—3.5 to 4.0—we would expect to see approximately 68% of the cases.

But what does this all mean you ask? If we are dealing with raw data and we know the mean and standard deviation of a sample, we can *predict* the intervals within which 68, 95 and 99% of our cases would be expected to fall. We call these intervals the—guess what—68, 95 and 99% confidence intervals. Now, here's where everything should come together in one great aha! experience if you've been following along.

If we had a *sampling distribution*, we would be able to predict the 68, 95 and 99% confidence intervals for where the population parameter should be! And isn't that why we sampled in the first place? So that we could predict where the population is on that variable? There's only one hitch. We don't actually have the sampling distribution! But we do have the distribution for the sample itself.

And we can from that distribution estimate the standard error (the sampling error) because it is based on the standard deviation and we have that. And, of course, we don't actually know the population parameter value—we're trying to find that out—but we can use our best estimate for that—the sample statistic. Now, if we have the mean of the sampling distribution (or set it to the mean from our sample) and we have an estimate of the standard error (we calculate that from our sample) then we have the two key ingredients that we need for our sampling distribution in order to estimate confidence intervals for the population parameter. Perhaps an example will help. Let's assume we did a study and drew a single sample from the population.

Furthermore, let's assume that the average for the sample was 3.75 and the standard deviation was 25. Now, what would the sampling distribution be in this case? Well, we don't actually construct it (because we would need to take an infinite number of samples) but we *can* estimate it. For starters, we assume that the mean of the sampling distribution is the mean of the sample, which is 3.75. Then, we calculate the standard error. To do this, we use the standard deviation for our sample and the sample size (in this case N=100) and we come up with a standard error of.025 (just trust me on this). Now we have everything we need to estimate a confidence interval for the population parameter. We would estimate that the probability is 68% that the true parameter value falls between 3.725 and 3.775 (*i.e,* 3.75 plus and minus.025); that the 95% confidence interval is 3.700 to 3.800; and that we can say with 99% confidence that the population value is between 3.675 and 3.825. The real value (in this fictitious example) was 3.72 and so we have correctly estimated that value with our sample.

TYPES OF SAMPLING

We may then consider different types of probability samples. Although there are a number of different methods that might be used to create a sample, they generally can be grouped into one of two categories: *probability samples* or *non-probability* samples.

Probability Samples

The idea behind this type is random selection. More specifically, each sample from the population of interest has a known probability of selection under a given sampling plan..

Simple Random Sampling

The most widely known type of a random sample is the simple random sample (SRS). This is characterized by the fact that the probability of selection is the same for every case in the population. Simple random sampling is a method of selecting n units from a population of size N such that every possible sample of size an has equal chance of being drawn.

An example may make this easier to understand. Imagine you want to carry out a survey of 100 voters in a small town with a population of 1,000 eligible voters. With a town this size, there are "old-fashioned" ways to draw a sample. For example, we could write the names of all voters on a piece of paper, put all pieces of paper into a box and draw 100 tickets at random. You shake the box, draw a piece of paper and set it aside, shake again, draw another, set it aside, etc. until we had 100 slips of paper. These 100 form our sample. And this sample would be drawn through a simple random sampling procedure - at each draw, every name in the box had the same probability of being chosen.

In real-world social research, designs that employ simple random sampling are difficult to come by. We can imagine some situations where it might be possible - you want to interview a sample of doctors in a hospital about work conditions. So you get a list of all the physicians that work in the hospital, write their names on a piece of paper, put those pieces of paper in the box, shake and draw.

But in most real-world instances it is impossible to list everything on a piece of paper and put it in a box, then randomly draw numbers until desired sample size is reached. There are many reasons why one would choose a different type of probability sample in practice.

Stratified Random Sampling

In this form of sampling, the population is first divided into two or more mutually exclusive segments based on some categories of variables of interest in the research. It is designed to organize the population into homogenous subsets before sampling, then drawing a random sample within each subset. With stratified random sampling the population of N units is divided into subpopulations of units respectively.

These subpopulations, called *strata*, are non-overlapping and together they comprise the whole of the population. When these have been determined, a sample is drawn from each, with a separate draw for each of the different strata. The sample sizes within the strata are denoted by respectively. If a SRS is taken within each stratum, then the whole sampling procedure is described as stratified random sampling.

The primary benefit of this method is to ensure that cases from smaller strata of the population are included in sufficient numbers to allow comparison. An example makes it easier to understand. Say that you're interested in how job satisfaction varies by race among a group of employees at a firm. To explore this issue, we need to create a sample of the employees of the firm. *However, the employee population at this particular firm is predominantly white*:

- *African-American*: 9%
- *Asian*: 10%
- *Latino:* 6%
- *White:* 75%

If we were to take a simple random sample of employees, there's a good chance that we would end up with very small numbers of Blacks, Asians, and Latinos. That could be disastrous for our research, since we might end up with too few cases for comparison in one or more of the smaller groups.

Rather than taking a simple random sample from the firm's population at large, in a stratified sampling design, we ensure that appropriate numbers of elements are drawn from each racial group in proportion to the percentage of the population as a whole. Say we want a sample of 1000 employees - we would stratify the sample by race then randomly draw out 750 employees from the White group, 90 from the African American, 100 from the Asian, and 60 from the Latino. This yields a sample that is proportionately representative of the firm as a whole. Stratification is a common technique.

There are many reasons for this, such as:

- If data of known precision are wanted for certain subpopulations, than each of these should be treated as a population in its own right.
- Administrative convenience may dictate the use of stratification, for example, if an agency administering a survey may have regional offices, which can supervise the survey for a part of the population.
- Sampling problems may be inherent with certain sub populations, such as people living in institutions.
- Stratification may improve the estimates of characteristics of the whole population. It may be possible to divide a heterogeneous population into sub-populations, each of which is internally homogenous. If these strata are homogenous, *i.e.*, the measurements vary little from one part to another; a precise estimate of any stratum mean can be obtained from a small sample in that stratum. The estimate can then be combined into a precise estimate for the whole population.
- There is also a statistical advantage in the method, as a stratified random sample nearly always results in a smaller variance for the estimated mean or other population parametres of interest.

Systematic Sampling

This method of sampling is at first glance very different from SRS. In practice, it is a variant of simple random sampling that involves some listing of elements - every nth element of list is then drawn for inclusion in the sample. Say you have a list of 10,000 people and you want a sample of 1,000.

Creating such a sample includes three steps:

1. Divide number of cases in the population by the desired sample size. In this example, dividing 10,000 by 1,000 gives a value of 10.
2. Select a random number between one and the value attained in Step 1. In this example, we choose a number between 1 and 10 - say we pick 7.
3. Starting with case number chosen in Step 2, take every tenth record.

More generally, suppose that the N units in the population are ranked 1 to N in some order. To select a sample of n units, we take a part at random, from the 1st k units and take every k-th part thereafter.

The advantages of systematic sampling method over simple random sampling include:

1. It is easier to draw a sample and often easier to execute without mistakes. This is a particular advantage when the drawing is done in the field.
2. Intuitively, you might think that systematic sampling might be more precise than SRS. In effect it stratifies the population into n strata, consisting of the 1st k units, the 2nd k units, and so on. Thus, we might expect the systematic sample to be as precise as a stratified random sample with one part per stratum. The difference is that with the systematic one the units occur at the same relative position in the stratum whereas with the stratified, the position in the stratum is determined separately by randomization within each stratum.

REPRESENTATIVE SAMPLING

A sample is *representative* when it is an accurate proportional representation of the population under study. If you want to study the attitudes of UT students regarding library services, it would not be enough to interview every 100th person who walked into the library. That technique would only measure the attitudes of UT students who use the library, not those who do not. In addition, it would only measure the attitudes of UT students who happened to use the library during the time you were collecting data. Therefore, the sample would not be very representative of UT students in general. In order to be a truly representative sample, every student at UT would have to have had an equal chance of being chosen to participate in the survey. This is called *randomization*.

If you stood in front of the student union and walked up to students, asking them questions, you still would not have a random sample. You would only be questioning students who happened to come to campus that day, and further, those that happened to walk past the student union. Those students who never walk that way would have had no chance of being questioned. In addition, you might unintentionally be biased as to who you question. You might unconsciously choose not to question students who look preoccupied or busy, or students who don't look like friendly people. This would invalidate your results, since your sample would not be randomly selected.

If you took a list of UT students, uploaded it onto a computer, then instructed the computer to randomly generate a list of 2 percent of all UT students, then your sample still might not be representative. What if, purely by chance, the computer did not include the correct proportion of seniors, or honours students, or graduate students? In order to further ensure that the sample is truly representative of the population, you might want to use a sampling technique called *stratification*. In order to stratify a population, you need to decide what sub-categories of the population might be statistically significant. For instance, graduate students as a group probably have different opinions than undergraduates regarding library usage, so they should be recognized as separate strata of the population. Once you have a list of the different strata, along with their respective percentages, you could instruct the computer to again randomly select students, this time taking care that a certain percentage are graduate students, a certain percentage are honours students, and a certain percentage are seniors. You would then come up with a more truly representative sample.

PROBABILITY AND NONPROBABILITY SAMPLING

A probability sampling scheme is one in which every unit in the population has a chance (greater than zero) of being selected in the sample, and this probability can be accurately determined. The combination of these traits makes it possible to produce unbiased estimates of population totals, by weighting sampled units according to their probability of selection.

Example: We want to estimate the total income of adults living in a given street. We visit each household in that street, identify all adults living there, and randomly select one adult from each household. (For example, we can allocate each person a random number, generated from a uniform distribution between 0 and 1, and select the person with the highest number in each household). We then interview the selected person and find their income. People living on their own are certain to be selected, so we simply add their income to our estimate of the total. But a person living in a household of two adults has only a one-in-two chance of selection. To reflect this, when we come to such a household, we would count the selected person's income twice towards the total. (In effect, the person who is selected from that household is taken as representing the person who isn't selected.)

In the above example, not everybody has the same probability of selection; what makes it a probability sample is the fact that each person's probability is known. When every element in the population *does* have the same probability of selection, this is known as an 'equal probability of selection' (EPS) design. Such designs are also referred to as 'self-weighting' because all sampled units are given the same weight. Probability sampling includes: Simple Random Sampling, Systematic Sampling, Stratified Sampling, Probability Proportional to Size Sampling, and Cluster or Multistage Sampling. These various ways of probability sampling have two things in common:

1. Every element has a known nonzero probability of being sampled and
2. involves random selection at some point.

Nonprobability sampling is any sampling method where some elements of the population have *no* chance of selection (these are sometimes referred to as 'out of coverage'/'undercovered'), or where the probability of selection can't be accurately determined.

It involves the selection of elements based on assumptions regarding the population of interest, which forms the criteria for selection. Hence, because the selection of elements is nonrandom, nonprobability sampling does not allow the estimation of sampling errors. These conditions give rise to exclusion bias, placing limits on how much information a sample can provide about the population. Information about the relationship between sample and population is limited, making it difficult to extrapolate from the sample to the population.

Example: We visit every household in a given street, and interview the first person to answer the door. In any household with more than one occupant, this is a nonprobability sample, because some people are more likely to answer the door (e.g. an unemployed person who spends most of their time at home is more likely to answer than an employed housemate who might be at work when the interviewer calls) and it's not practical to calculate these probabilities.

Nonprobability Sampling includes: Accidental Sampling, Quota Sampling and Purposive Sampling. In addition, nonresponse effects may turn *any* probability design into a nonprobability design if the characteristics of nonresponse are not well understood, since nonresponse effectively modifies each element's probability of being sampled.

Sampling Methods

Within any of the types of frame identified above, a variety of sampling methods can be employed, individually or in combination. Factors commonly influencing the choice between these designs include:

- Nature and quality of the frame
- Availability of auxiliary information about units on the frame
- Accuracy requirements, and the need to measure accuracy
- Whether detailed analysis of the sample is expected
- Cost/operational concerns.

Simple Random Sampling

In a simple random sample ('SRS') of a given size, all such subsets of the frame are given an equal probability. Each element of the frame thus has an equal probability of selection: the frame is not subdivided or partitioned. Furthermore, any given *pair* of elements has the same chance of selection as any other such pair (and similarly for triples, and so on). This minimises bias and simplifies analysis of results. In particular, the variance between individual results within the sample is a good indicator of variance in the overall population, which makes it relatively easy to estimate the accuracy of results.

However, SRS can be vulnerable to sampling error because the randomness of the selection may result in a sample that doesn't reflect the makeup of the population. For instance, a simple random sample of ten people from a given country will *on average* produce five men and five women, but any given trial is likely to overrepresent one sex and underrepresent the other.

SRS may also be cumbersome and tedious when sampling from an unusually large target population. In some cases, investigators are interested in research questions specific to subgroups of the population. For example, researchers might be interested in examining whether cognitive ability as a predictor of job performance is equally applicable across racial groups. SRS cannot accommodate the needs of researchers in this situation because it does not provide subsamples of the population. Simple random sampling is always an EPS design, but not all EPS designs are simple random sampling.

Systematic Sampling

Systematic sampling relies on arranging the target population according to some ordering scheme and then selecting elements at regular intervals through that ordered list. Systematic sampling involves a random start and then proceeds with the selection of every *k*th element from then onwards. In this case, *k*=(population size/sample size).

It is important that the starting point is not automatically the first in the list, but is instead randomly chosen from within the first to the *k*th element in the list. A simple example would be to select every 10th name from the telephone directory (an 'every 10th' sample, also referred to as 'sampling with a skip of 10').

As long as the starting point is randomized, systematic sampling is a type of probability sampling. It is easy to implement and the stratification induced can make it efficient, *if* the variable by which the list is ordered is correlated with the variable of interest. 'Every 10th' sampling is especially useful for efficient sampling from databases.

Example: Suppose we wish to sample people from a long street that starts in a poor district (house #1) and ends in an expensive district (house #1000). A simple random selection of addresses from this street could easily end up with too many from the high end and too few from the low end (or vice versa), leading to an unrepresentative sample. Selecting (e.g.) every 10th street number along the street ensures that the sample is spread evenly along the length of the street, representing all of these districts. (Note that if we always start at house #1 and end at #991, the sample is slightly biased towards the low end; by randomly selecting the start between #1 and #10, this bias is eliminated.)

However, systematic sampling is especially vulnerable to periodicities in the list. If periodicity is present and the period is a multiple or factor of the interval used, the sample is especially likely to be *un*representative of the overall population, making the scheme less accurate than simple random sampling.

Example: Consider a street where the odd-numbered houses are all on the north (expensive) side of the road, and the even-numbered houses are all on the south (cheap) side. Under the sampling scheme given above, it is impossible' to get a representative sample; either the houses sampled will all be from the odd-numbered, expensive side, or they will all be from the even-numbered, cheap side.

Another drawback of systematic sampling is that even in scenarios where it is more accurate than SRS, its theoretical properties make it difficult to *quantify* that accuracy. (In the two examples of systematic sampling that are given above, much of the potential sampling error is due to variation between neighbouring houses-but because this method never selects two neighbouring houses, the sample will not give us any information on that variation.)

As described above, systematic sampling is an EPS method, because all elements have the same probability of selection (in the example given, one in ten). It is *not* 'simple random sampling' because different subsets of the same size have different selection probabilities-e.g. the set {4,14,24,...,994} has a one-in-ten probability of selection, but the set {4,13,24,34,...} has zero probability of selection.

Non-probability Sampling

Social research is often conducted in situations where a researcher cannot select the kinds of probability samples used in large-scale social surveys. For example, say you wanted to study homelessness - there is no list of homeless individuals nor are you likely to create such a list. However, you need to get some kind of a sample of respondents in order to conduct your research. To gather such a sample, you would likely use some form of non-probability sampling.

To reiterate, the primary difference between probability methods of sampling and non-probability methods is that in the latter you do not know the likelihood that any element of a population will be selected for study. There are four primary types of non-probability sampling methods:

Availability Sampling

Availability sampling is a method of choosing subjects who are available or easy to find. This method is also sometimes referred to as haphazard, accidental, or convenience sampling. The primary advantage of the method is that it is very easy to carry out, relative to other methods. A researcher can merely stand out on his/her favourite street corner or in his/her favourite tavern and hand out surveys. One place this used to show up often is in university courses. Years ago, researchers often would conduct surveys of students in their large lecture courses.

For example, all students taking introductory sociology courses would have been given a survey and compelled to fill it out. There are some advantages to this design - it is easy to do, particularly with a captive audience, and in some schools you can attain a large number of interviews through this method.

The primary problem with availability sampling is that you can never be certain what population the participants in the study represent. The population is unknown, the method for selecting cases is haphazard, and the cases studied probably don't represent any population you could come up with.

However, there are some situations in which this kind of design has advantages - for example, survey designers often want to have some people respond to their survey before it is given out in the "real" research setting as a way of making certain the questions make sense to respondents. For this purpose, availability sampling is not a bad way to get a group to take a survey, though in this case researchers care less about the specific responses given than whether the instrument is confusing or makes people feel bad.

Despite the known flaws with this design, it's remarkably common. Ask a provocative question, give telephone number and web site address, announce results of poll. This method provides some form of statistical data on a current issue, but it is entirely unknown what population the results of such polls represents. At best, a researcher could make some cond-itional statement about people who are watching CNN at a particular point in time who cared enough about the issue in question to log on or call in.

Quota Sampling

Quota sampling is designed to overcome the most obvious flaw of availability sampling. Rather than taking just anyone, you set quotas to ensure that the sample you get represents certain characteristics in proportion to their prevalence in the population.

Note that for this method, you have to know something about the characteristics of the population ahead of time. Say you want to make sure you have a sample proportional to the population in terms of gender - you have to know what percentage of the population is male and female, then collect sample until yours matches. Marketing studies are particularly fond of this form of research design.

The primary problem with this form of sampling is that even when we know that a quota sample is representative of the particular characteristics for which quotas have been set, we have no way of knowing if sample is representative in terms of any other characteristics. If we set quotas for gender and age, we are likely to attain a sample with good representativeness on age and gender, but one that may not be very representative in terms of income and education or other factors.

Moreover, because researchers can set quotas for only a small fraction of the characteristics relevant to a study quota sampling is really not much better than availability sampling. To reiterate, you must know the characteristics of the entire population to set quotas; otherwise there's not much point to setting up quotas.

Finally, interviewers often introduce bias when allowed to self-select respondents, which is usually the case in this form of research. In choosing males 18-25, interviewers are more likely to choose those that are better-dressed, seem more approachable or less threatening. That may be understandable from a practical point of view, but it introduces bias into research findings.

Purposive Sampling

Purposive sampling is a sampling method in which elements are chosen based on purpose of the study. Purposive sampling may involve studying the entire population of some limited group or a subset of a population. As with other non-probability sampling methods, purposive sampling does not produce a sample that is representative of a larger population, but it can be exactly what is needed in some cases - study of organization, community, or some other clearly defined and relatively limited group.

Snowball Sampling

Snowball sampling is a method in which a researcher identifies one member of some population of interest, speaks to him/her, then asks that person to identify others in the population that the researcher might speak to. This person is then asked to refer the researcher to yet another person, and so on. Snowball sampling is very good for cases where members of a special population are difficult to locate. For example, several studies of Mexican migrants in Los Angeles have used snowball sampling to get respondents.

The method also has an interesting application to group membership - if you want to look at pattern of recruitment to a community organization over time, you might begin by interviewing fairly recent recruits, asking them who introduced them to the group. Then interview the people named, asking them who recruited them to the group.

The method creates a sample with questionable representativeness. A researcher is not sure who is in the sample. In effect snowball sampling often leads the researcher into a realm he/she knows little about. It can be difficult to determine how a sample compares to a larger population. Also, there's an issue of who respondents refer you to - friends refer to friends, less likely to refer to ones they don't like, fear, etc.

APPLICATIONS OF STATISTICS

Sociology is one of the social sciences aiming to discover the basic structure of human society, to identify the main forces that hold groups together or weaken them and to learn the conditions that transform social life. It highlights and illuminates aspects of social life that otherwise might be only obscurely recognized and understood.

The sociologist may be called upon for help with a special problem such as social conflict, urban plight or the war on poverty or crimes. His practical contribution lies in the ability to clarify the underlaying nature of social problems to estimate more exactly their dimensions and to identify aspects that seem most amenable to remedy with the knowledge and skills at hand.

He naturally lands in sociological research which is the purposeful effort to learn more about society than one can in the ordinary course of living. Keeping in view of the problem he sets forth his objectives collects materials or data and uses statistical techniques and the knowledge and theory already established on similar topics to achieve his objectives.

So statistical data and statistical methods are quite indispensable for sociological research studies. There is a growing emphasis recently on social survey methods or research methodology in all faculties of arts.

Sociologists seek the help of statistical tools to study cultural change in the society, family pattern, prostitution, crime, marriage system etc. They also study statistically the relation between prostitution and poverty, crime and poverty, drunkness and crime, illiteracy and crime etc. Thus statistics is of immense use in various sociological studies.

Statistics and Government

The functions of a government are more varied and complex. Various depts in the state are required to collect and record statistical data in a systematic manner for an effective administration. Data pertaining to various fields namely population, natural resources, production both agricultural and industrial, finance, trade, exports and imports, prices, labour, transport and communication, health, education, defence, crimes etc are the most fundamental requirements of the state for its administration.

It is only on this basis of such data; the government decides on the priority areas, gives more attention to them through target oriented programmes and studies the impact of the programmes for its future guidelines.

Statistics and Planning

Modern age is an age of planning and statistics are indispensable for planning. According to Tippett planning greater or lesser degree according to the government in power is the order of the day and without statistics, planning is inconceivable. Based only on a correct assessment of various resources both human and material of the country proper planning can be made. A study of data relating to population, agriculture, industry, prices, employment, health, education enables the planners to fix up time-bound targets on the social and economic fronts evaluation of such economic and social programmes at different stages by means of related data gathered continuously and systematically is also done to decide whether the programmes are on towards the goal or targets set.

Statistics and Economics

In the fields of economics it is almost impossible to think of a problem which does not require an extensive use of statistical data. Most of the laws in economics are based on a study of a large number of units and their analysis is enabled by statistical data and the statistical methods. The important economic aspects like production, consumption, exchange and distribution are described, compared and correlated with the aid of statistical tools. By a statistical study of time series on prices, sales, production one can study their trends, fluctuations and the underlaying causes. Thus statistics is indispensable in economic analysis.

METHODS OF CENTRAL TENDENCY OF AVERAGES

Condensation of data is necessary for a proper statistical analysis. A large number of big numbers are not only confusing to mind but also difficult to analyse. After a thorough scrutiny of collected data, classification which is a process of arranging data into different homogenous classes according to resemblances and similarities is carried out first. Then of course tabulation of data is resorted to. The classification and tabulation of the collected data besides removing the complexity render condensation and comparison. An average is defined as a value which should represent the whole mass of data. It is a typical or central value summarizing the whole data. It is also called a measure of central tendency for the reason that the individual values in the data show some tendency to centre about this average. It will be located in between the minimum and the maximum of the values in the data.

There are five types of average which are;

1. Arithmatic Mean
2. Median

3. Mode
4. Geometric Mean and
5. Harmonic Mean.

Arithmetic Mean

The Arithmetic mean or simply the mean is the best known easily understood and most frequently used average in any statistical analysis. It is defined as the sum of all the values in the data.

Median

Median is another widely known and frequently used average. It is defined as the most central or the middle most value of the data given in the form of an array. By an array, we mean an arrangement of the data either in ascending order or descending order of magnitude. In the case of ungrouped data one has to form an array first and then locate the middle most value which is the median.

Relative Merits and Demerits of Mean, Median and Mode

Mean: The mean is the most commonly and frequently used average. It is a simple average, understandable even to a layman. It is based on all the values in a given data. It is easy to calculate and is basic to the calculation of further statistical measures of dispersion, correlation etc. Of all the averages, it is the most stable one. However it has some demerits.

It gives undue weightages to extreme value. In other words it is greatly influenced by extreme values. Moreover; it cannot be calculated for data with open-ended classes at the extreme. It cannot be fixed graphically unlike the median or the mode. It is the most useful average of analysis when the analysis is made with full reference to the nature of individual values of the data. Inspite of a few shortcomings; it is the most satisfactory average.

Median: The median is another well-known and widely used average. It is well-defined formula and is easily understood. It is advantageously used as a representative value of such factors or qualities which cannot be measured. Unlike the mean, median can be located graphically. It is also possible to find the median for data with open ended classes at the extreme. It is amenable for further algebraic processes. However, it is an average, not based on all the values of the given data. It is not as stable as the mean. It has only a limited use in practice.

Mode: It is a useful measure of central tendency, as a representative of the majority of values in the data. It is a practical average, easily understood by even laymen. Its calculations are not difficult. It can be ascertained even for data with open-ended classes at the extreme. It can be located by graphical means using a frequency curve. The mode is not based on all the values in the data. It become less useful when the data distribution is not unit-model. Of all the averages, it is the most unstable average.

MEASURES OF DISPERSION

When a mass of quantitative data is collected for a statistical purpose, it is tabulated to a form with in view that its characteristics as a whole may be readily determined. A single significant and representative expression or a measure called average is then computed to summarize or explain as a whole the entire data.

This form of a comparing two or more groups or series but an average along is not a satisfactory criterion for such a purpose. Dispersion means the extent of values around some average. There are four measures of dispersion, namely

- Range
- Quartile Deviation or semi interquartile range
- Mean Deviation
- Standard Deviation.

Range

Range is the simplest measure of dispersion. It is defined as the positive difference between the largest and the smallest values in the given data. It is easily understood and computed but depends exclusively on the two extreme values while it is desirable to have a measure dependent on all the values. The range is a very useful measure in statistical quality control of products in industries wherein the interest lies in getting a quick rather than an accurate measure of variability.

Quartile Deviation or Semi-interquartile Range

This measure is based on two measures called the lower or first quartile and the upper or third quartile. The lower quartile denoted as Q1 is defined as the value which leaves ¼ of the values below it when the data forms an array and the third quartile denoted as Q3 is the value which leaves 3/4s of values below it when the data forms an array. Once Q1 and Q3 are known, the quartile deviation Q-D is given by; Q-D= Q3-Q1/2 and Q3-Q1 is called the interquartile range

Mean Deviation

The mean deviation of a data is defined as the means of the absolute deviation of values from some average especially the arithmetic mean or median. It is a better measure of dispersion than range and Q D as it takes into account, all the values in the given data for ungrouped data. M. D about median M can also be defined by considering median instead of mean.

Standard Deviation

The standard deviation abbreviated as S. D and symbolically represents as sigma is the most important and wide used measure of dispersion. It is defined as the root mean square deviation of values from their mean i.e. it is the square root of the means of square deviation of values from their mean. The square of the s. d. is called the variance.

7

Organizational Research and Management

FUNCTIONS OF MANAGEMENT

Management operates through various functions, often classified as planning, organizing, leading/motivating and controlling.

- *Planning:* deciding what has to happen in the future (today, next week, next month, next year, over the next five years, etc.) and generating plans for action.
- *Organizing:* making optimum use of the resources required to enable the successful carrying out of plans.
- *Leading/Motivating:* exhibiting skills in these areas for getting others to play an effective part in achieving plans.
- *Controlling:* monitoring-checking progress against plans, which may need modification based on feedback.

Management of a business concern is a highly complex phenomenon and it is not possible to consider all its facets at the same moment. In the process of managing, a manager performs a variety of functions to accomplish the pre-determined goals. The authorities and the scholars of management differ widely about managerial functions and their classification. A widely accepted classification was given quite early by Henri Fayol. He described them as follows:

(i) forecasting and planning,
(ii) organising,
(iii) commanding,
(iv) co-ordinating, and
(v) controlling.

Later, several authorities have been adding other functions to the list. Dale and Michelon have described them as planning, organising, staffing, directing, controlling, innovation, presentation, communication. Greenwood suggests planning, organising, staffing, directing, controlling and decision-making. Massies states them as decision-making, planning, organising, staffing, controlling, and communicating. Alien describes them as planning, co-ordinating, motivating, communicating and organising. According to Terry, they are: planning, organising, staffing, actuating, and controlling. According to Koontz and O'Donnell, "the most useful method of classifying managerial functions is to group them around the activities of planning, organising, staffing, directing and controlling".

However, according to them, in practice it is not always possible "to slice all managerial activities neatly into these categories, since the functions tend to coalesce." Similar views have been expressed by C.S. George when he states that: "The management process, therefore, is not a series of separate functions (planning, organising, directing, controlling) which can be performed independently; it is a composite process made up of these individual ingredients. No one function can be performed without involving the others, and by employing his composite, managers create an environment conducive to the co-ordinated participation of team members."

Moreover, a distinction has to be made between management functions and the operating tasks or business functions it performs, i.e., manufacturing, purchasing, selling, financing, accounting etc. Normally and theoretically there may be some sequential order in which these functions are performed. In other words, planning comes first and organising, staffing, directing, co-ordinating and controlling follow it. But each of these functions is inter-linked and interwoven in actual practice, and a manager need not follow this order. He solves the problems as and when they arise without giving much attention to this order. He can move easily from one 'function to another and devote attention to the most pressing problem before him. It has been rightly pointed out that "Managing is, after all, a systematic network and not a sequentially undertaken set of duties."

We must recognise that there is a Principle of Universality of Managerial Functions. Management at different levels—top, middle, or lower—perform the same set of functions. Similarly, these very functions have to be performed in every type of undertaking, business or non-business, small or large, trading or manufacturing, philanthropic or profit-making, partnership or joint stock company, university, hotel, club, military or Government departments. As managers, they all are engaged in getting things done through others and are performing the same set of functions. According to Koontz and O'Donnell, this principle has many implications, but two of them are very important. *Firstly*, it means that "anything significant that is said about the functions of one manager applies to all managers". *Secondly*, it implies that "managerial knowledge and experience are transferable from department to department and from enterprise to enterprise; merchandising executives may be shifted to manufacturing; or the military commander to peaceful pursuits: or foremen from flour milling to warehousing; and production managers to sales."

Having discussed the nature and classification of various functions of management, let us now discuss each of these functions, namely, planning, organising, staffing, directing, coordinating and controlling, in detail. Planning is the most basic and important of all management functions and precedes all other functions or management activities. Every manager plans before he acts.

According to George R. Terry, it is "the selecting and relating of facts in the visualization and formation of proposed activities believed necessary to achieve desired results". In the words of Koontz and O'Donnell, "Plans involve selecting enterprise objectives and departmental goals and determining ways of reaching them. Plans thus provide a rational approach to pre-selected objectives." They further observe that "Planning is deciding in advance what to do, when to do it, and who is to do it. Planning bridges the gap from where we are to where we want to go. It makes it possible for things to occur which would not otherwise happen."

Planning thus is intellectual or mental process to foresee the future events and decide about the future course of actions and the manner in which they have to be taken. It necessarily involves thinking about; different lines of actions, and would imply looking ahead and conceptualizing the future events and deciding in advance the various alternatives. Of course, the plan has to be linked with the objectives of the firm. Since there may be several possibilities, a series of plans may have to be developed. But all these plans need not be implemented. However, the management must be ready, in advance, to tackle all the future possibilities.

A plan may be formal or informal. It is formal when it is reduced to writing. It would be advantageous to prepare a formal plan specially when a large number of actions has to be taken in future.

This is necessary to avoid oversight or inattention by the management. An informal plan is one which is not reduced to writing but is kept only in the mind of the manager. If the number of actions to be taken is less and the actions have to be taken in a short period informal planning may be usefully adopted. Plans may be short-term or long-term depending upon the period which is kept in view while formulating a plan.

Similarly, there may be single-use or standing plans. A single-use plan is one which sets a course of action for a specific situation and is used up once the particular goal is reached, whereas the standing plan is one which is designed to be used over and over again.

Planning is to be done at all levels of organisation, but the lower-level management tends to derive its own plans from the plan of higher-level management. However, higher-level management spends more time and energy in planning than lower-level management. Higher-level management keeps broader objective of the company as a whole and takes a long-run view whereas others keep in mind only immediate objective and sectional approach in planning.

BASIC METHODS OF MARKET RESEARCH

While there are many ways to perform market research, most businesses use one or more of five basic methods: surveys, focus groups, personal interviews, observation, and field trials. The type of data you need and how much money you're willing to spend will determine which techniques you choose for your business.

Surveys: With concise and straightforward questionnaires, you can analyze a sample group that represents your target market. The larger the sample, the more reliable your results will be.

- In-person surveys are one-on-one interviews typically conducted in high-traffic locations such as shopping malls. They allow you to present people with samples of products, packaging, or advertising and gather immediate feedback. In-person surveys can generate response rates of more than 90 percent, but they are costly. With the time and labour involved, the tab for an in-person survey can run as high as $100 per interview.
- Telephone surveys are less expensive than in-person surveys, but costlier than mail. However, due to consumer resistance to relentless telemarketing, convincing people to participate in phone surveys has grown increasingly difficult. Telephone surveys generally yield response rates of 50 to 60 percent.
- Mail surveys are a relatively inexpensive way to reach a broad audience. They're much cheaper than in-person and phone surveys, but they only generate response rates of 3 percent to 15 percent. Despite the low return, mail surveys remain a cost-effective choice for small businesses.
- Online surveys usually generate unpredictable response rates and unreliable data, because you have no control over the pool of respondents. But an online survey is a simple, inexpensive way to collect anecdotal evidence and gather customer opinions and preferences.

Focus groups: In focus groups, a moderator uses a scripted series of questions or topics to lead a discussion among a group of people. These sessions take place at neutral locations, usually at facilities with videotaping equipment and an observation room with one-way mirrors. A focus group usually lasts one to two hours, and it takes at least three groups to get balanced results.

Personal interviews: Like focus groups, personal interviews include unstructured, open-ended questions. They usually last for about an hour and are typically recorded. Focus groups and personal interviews provide more subjective data than surveys. The results are not statistically reliable, which means that they usually don't represent a large enough segment of the population. Nevertheless, focus groups and interviews yield valuable insights into customer attitudes and are excellent ways to uncover issues related to new products or service development.

Observation: Individual responses to surveys and focus groups are sometimes at odds with people's actual behaviour. When you observe consumers in action by videotaping them in stores, at work, or at home, you can observe how they buy or use a product. This gives you a more accurate picture of customers' usage habits and shopping patterns.

Field trials: Placing a new product in selected stores to test customer response under real-life selling conditions can help you make product modifications, adjust prices, or improve packaging. Small business owners should try to establish rapport with local store owners and Web sites that can help them test their products.

MATCHING QUALITATIVE RESEARCH TO THE RESEARCH OBJECTIVES

Qualitative research can be used as an *exploratory* tool. Where there is uncertainty about a subject, and the researcher does not know what detailed questions to ask, a few depth interviews or groups may be sufficient to provide an understanding and explanations which answer the problem. Imagine that you are a manager of a hospital, eager to improve the standard of patient care.

Your 'clients' include the young, the mentally ill, the old and the very poorly. What they think of your service is decisive in guiding how you develop in the future – but how do you find out what they think?

The very young, the very old and the mentally ill may be unable to tell you. And when do you interview them? If you do so while they are in the hospital, it could bias the result as their treatment may not yet be over and, in any case, they may find it hard to be critical during the period of medical care. It may be better to interview soon after they return home, even though this is the more expensive option.

Think About

The next time you talk to someone in business, pay attention to their body language. What does it tell you? What other clues surrounding the person tell you about them such as the books on their shelves, the watch they are wearing, their clothes, and the way they organise themselves? How much more can you learn about this person and their attitudes from paying attention to non verbal leakage?

Designing a quantitative study for checking on standards of patient care is likely to be complicated and costly. Instead of focusing on the patients themselves, an understanding of the issues could be obtained from people who can speak on their behalf. Group discussions with general practitioners in the hospital's territory could provide a distilled view of the strengths and weaknesses of the hospital and individual departments. Group discussions with GPs could be quickly convened and the results available within a couple of weeks.

A couple of focus groups is not a sizeable study but they could provide a signpost as to where the real problems lie and give focus to any subsequent quantitative research.

In the same vein, researchers may be faced with a marketing problem which needs investigation. Problem solving requires an understanding of the cause and effects and these could be flushed out by discussion in groups or depths. When used in this capacity, qualitative research is a *diagnostic* tool. Typical subjects for analysis could be a downturn in sales, a loss of market share or an increase in complaints. A major UK manufacturer of domestic heating boilers looked at the latest trade association figures and saw that it had suffered a loss of share. This had not been apparent from the company's sales figures which were holding up quite well in a market which was buoyant. Four group discussions with plumbers who install boilers were sufficient to show that a major competitor was heavily discounting and using direct mail to promote its boilers – a method of promotion which was concealed from the competing suppliers.

Very often, qualitative research is used for *creative* inspiration and guidance. For example, if an advertising agency wants ideas for a new campaign it may get these from the interaction of thoughts arising in focus groups. The creative power of qualitative research need not be confined to teasing out ideas for advertising; it can be used to guide all types of innovative work such as new product development and branding.

A manufacturer of crisps was concerned that the health food trend would eventually affect his market and wanted to know in which way he could respond. Qualitative research was commissioned to find out if, with some modification, crisps could be positioned as a health food. Qualitative research can also be used to *evaluate* ideas. Focus groups or depth interviews can be used to find out what people think of a different presentation of an advert or pack design. Of course, there will be no large numbers to substantiate the evaluation but the views of the small number of respondents may be sufficiently convincing.

A cable manufacturer wanted to steal a march in the competitive field of house wiring and designed a pack which looked very different to the conventional reels which electricians had used for years.

The pack dispensed the cable from the centre of the reel and so offered a number of advantages. However, electricians are very much driven by habit and group discussions showed if there would be any resistance to the new product and how to overcome it.

FRAMEWORK FOR CHOOSING A RESEARCH DESIGN

One reason why secondary research sources are often omitted is institutional. Most research is carried out by market research agencies which for various reasons find problems in offering desk research as a profit earning service. In this respect in-house research can have an edge over buying-in.

Once desk research is completed (and assuming it does not yield all the information required for the project) primary research or fieldwork can be planned to fill in the gaps.

Think About

When did you last visit a commercial library? What do you think you could obtain from a library that you cannot obtain from the internet? Are you aware that some libraries will carry out short desk research assignments for you for a small fee? If you had £5,000 to spend, how much could you find out by desk research?

How much could you find out by primary research?

Decisions in primary research design.

When contemplating the research method for primary data, the following should be considered:

- Whether the nature of the information sought is primarily quantitative or qualitative.

Sampling issues including:

- The universe to be covered e.g.: all adults, housewives, buyers of specific products in consumer markets and comparable aspects of industrial markets.
- Sectors or sub-groupings of the whole universe which are of specific interest and need to be considered in sampling design.

- Sampling method and size (number of interviews).
- Data collection methods – e.g. face to face or phone interviews, postal survey, observation etc.
- How the resulting data will be analysed – particularly relevant for more sophisticated techniques such as conjoint analysis.

The basis of which of the above should be covered in the design, depends on the research objectives, the information coverage, and the accuracy sought. Or at least this should be so in theory. In practice other factors have to be taken into account and especially the resources available including the budget.

For example, the ideal method may be face to face interviews with a sample of 1,000 people in their home. However, but for practical reasons (speed and cost) street interviews with 200 may be used instead. There is, of course, some point beyond which compromises are a danger to the accuracy of the result. What budget should be made available for the research project?

This is not a question of arguing that the budget should be whatever is needed to meet the research objectives at the required accuracy level. It is more a question of what funds are available or can be afforded for the project relative to other calls on business expenditure. Furthermore, even if cash is freely available, there are other considerations and especially the amount at risk in the decision which the research is to guide. For example, a research budget of £30,000 may be well worth spending if it is to guide a decision that entails capital expenditure of £5 million. However, if the decision has much lower cost implications, the value of doing the research will be less and obviously there is no point to spending £30,000 on research to decide whether to invest in a project entailing only this level of expenditure. For this reason, many businesses cannot afford to test the effectiveness of their advertising campaigns because the cost of the research is more than the campaigns themselves.

Apart from a cash budget, the major resource required for a research project is suitably qualified people. An in-house researcher may be able to carry out desk research but would find it very difficult to carry out extensive fieldwork and for this would need the help of an agency.

A research plan needs a timetable. The deadline for the project may well be driven by external events such as a board meeting or input for a marketing plan.

The timeline available for the research often influences the research methods. Face to face interviews may be preferred but telephone interviews are quicker.

Certainly good research can be carried out within a short timetable but beyond some point, quality will be compromised if the timetable is squeezed too much.

The Case of the Softone Light Bulbs

At this stage the reader may find it helpful to think about a case study that incorporated a number of different design issues that have been discussed.

Philips Lighting is an established player in the light bulb market where it faced a number of challenges:

- Consumers are not very interested in the product
- They are sold at low prices
- They have a strong seasonality of sales (winter months) and low growth
- There is considerable competition and margins are low
- Light bulbs have low brand awareness and there is a strong own brand presence.

A product development from the Philips research laboratories was a technique for electrostatic coating the inside of a lamp to produce a softer light.

The additions of pigments offered the opportunity to produce light with a hint of colour. The new product was given the brand name *Softone.* Philips decided to use this product to stimulate interest in a commodity market and to help build its position as an innovative lighting supplier. Market research played a crucial role at all stages in the life cycle of this product development.

The first market research gateway was to establish that the product had significant consumer appeal, especially at a premium price. If the results of the research proved negative, it would be back to the technical drawing board to improve the product features or the launch would be dropped.

As is often the case with new product research, there is considerable pressure for the research to show a green light as the new product will have gained a momentum with many people in the organization banking on its success. The market researchers had to use a method that would deliver a true understanding of the new lamp's appeal and that would stand up to the political attacks it could face if the results were less than positive.

The research method chosen to test the new product's appeal was a survey of 200 face to face interviews with a quota sample of lamp buyers in two locations several hundred miles apart (Dewsbury in Yorkshire and Southampton in Hampshire). Respondents were recruited in shopping malls and brought into "halls". A sample of 200 could be expected to yield an accuracy of + or – 7% (ie if 50% of the sample expected to buy the lamp, the true level among the whole population might be expected to be between 43% and 57%). Restricting the sample to only two locations ruled out any true regional analysis but would at least indicate whether major geographical differences existed (it was assumed they did not).

Quotas were imposed to ensure that a mix of demographic groups representing the population as a whole was covered. In a hall test of this type and without quotas, there would be an over representation of lower social grades and older people.

The results from this preliminary research were positive and although not many people bought conventional coloured lamps, the interest in the pastel concept was high with 79% saying that they would be very or quite likely to buy the new product. A small price premium was acceptable to most people. It should be pointed out that it would be dangerous to extrapolate the high interest into a propensity to buy figure but it did suggest that there was a significant body of interest and it was worth going forward to the next stage.

Instead of carrying out the hall test, the research team could have opted for a more qualitative approach using focus groups. However, they believed that groups on their own would not have given them a feel for the size of the demand and some quantitative method would still have been needed. In the interests of budget and speed, the groups were skipped. Focus groups did play a part in the next phase of the research as the marketing team needed help with ideas for the pack design.

Philip's design agency developed various alternative pack concepts for evaluation and these were tested in four focus groups. Women aged 22 to 45 (the main group of lamp buyers) were recruited to the groups which were split between the North (two groups) and the South (two groups). It was also decided that the group respondents should be restricted to those with a strong interest in home decoration and colour coordination in their homes.

The focus groups confirmed the high level of interest in the new range of lamps and gave some interesting pointers for the pack designers. The brand *Softone* worked well on its own without the need to emphasise the words pastel or stronger colours. In fact, it was thought most appealing to communicate that the lamps offered just a hint of ....(colour) because subtlety in lighting was considered paramount. Philips backed the Softone launch with both television and press advertising.

This was going to incur a substantial sum and again, market research was called upon to help guide the decisions. Three alternative TV advertising styles were developed to the *animatics* stage (moving story boards) and the press adverts were developed in *storyboard* form (artist prepared visuals without the detail of the final artwork). Research had to establish the effectiveness of the alternative TV and press treatments in terms of impact, memorability and the interest in the Softone product.

It was thought that focus groups would not be a suitable method to test the advertising concepts as there could be contamination in respondents' views caused by the debate that would be sure to take place. It was considered better to carry out face to face depth interviews and for this the researchers used a *mini-hall test.*

Halls were taken in four locations and at each hall 40 people were shown the TV ads and 40 were shown the press ads. The interviews lasted half an hour and asked questions such as:

- What sort of product is being advertised? Is it a *new* product or just another coloured light bulb?
- What is the impact of the advert? Would it stop the reader/viewer and catch their attention?
- What is the message of the advert?
- Is the advert believable?
- Is the product for the respondents? Would they buy it?
- Where would it be used?

Having seen the adverts, the great majority of respondents were enthusiastic about Softone, they saw it as innovative and many were keen to have it in their homes. None of the adverts had a particularly good impact but there were some strong pointers as to how to make improvements. The more direct approaches in the adverts were considered important for a novel product where guidance in home use was thought to be needed. One of the TV ads and one of the press ads was stronger than the others and these were the ones that were modified by the agency for final use. (In theory the revised ads should have been tested again but it was essential not to delay the campaign any further so no further testing took place).

The campaign was extremely important to Philips and it was considered important that its effectiveness should be measured in terms of brand awareness, whether the product had been bought (and intentions to repurchase) and the recall of the content of the ads which were used. In the choice of the tracking design, the results from each stage of the research had to be capable of yielding awareness levels and other measures that showed a true difference and not one that could be within the bounds of sampling error. A sample size of 200 was considered the minimum in each of the 12 TV regions which meant that a national sample of 2,400 lamp buyers would need to be interviewed in each wave of the test. Philips knew from previous research that amongst a random sample of the adult population, two thirds would have bought a lamp within the last six months and therefore be classed as "lamp buyers" for the purposes of the survey. This meant that a minimum sample of 3,500 adults would have to be approached to achieve the 2,400 interviews.

The study aimed to track awareness and behaviour over a period of three years and this required nine stages – each of them with the large sample of

2,400 interviews. This would have been a very costly exercise if nine ad hoc studies had been carried out. Omnibus surveys offered the solution. Omnibus surveys run regularly, and subscribers can buy space on the questionnaire on a pay as you go basis.

Using this tracking measure, Philips was able to see that the recall of its Philips/Softone brand doubled from 15% to over 30% in the three years of the campaign.

Sometimes market research is used in an ad hoc way to guide a business decision. In the Philips Softone case study we saw how research was used continuously over a number of years to ensure the successful goal of building a significant share in the lighting market with a new and innovative brand of lamp.

The case study also illustrates that there are no methods or research designs that have to be used to meet an objective. Research designs are often a compromise between the trade-off of the accuracy of the information required against the time and cost of carrying out the research. We saw that Philips research team used a mixture of research designs:

- Hall tests to establish that the lamp concept was attractive and would be successful.
- Focus groups to get ideas for the pack design.
- Depth interviews to test TV and press ads.
- Omnibus interviews to track brand recall.

Market research followed the Softone product through a number of stages of its life cycle. In the Softone case study, there was very little secondary research used because Philips already had a good knowledge of the lighting industry. For other companies there may be a considerable use made of secondary research in the early stages of a product's life cycle.

EVOLUTION OF CLASSICAL APPROACH TO MANAGEMENT

Traditional process of learning is either through obsevation and experiment. Nature or environment is considered uniform and when we observe certain phenomenon or events uniformly leading to the same result or results, we conclude a cause and effect relationship between the two. This is learning by observation or in other words by experience. Earlier thinkers on management followed this approach in developing theories of management.

Learning principally is through emphirical process and through analysis of the data collected through observation. Draw the principles of managment by looking at and anyalysing the jobs that all managers commonly do. This approach served as a starting point for pioneers on management science to verify the validity and improve the applicability of the principles and practices of management. Analysis of observd data is what constitute a case study. The observational method of case study helps arriving at logical conclusions about past experience and to test the same as standards for future events.

The German sociolists, Max Weber followed the classical approach and developed his theory of Bureaucracy, which portrays the structure anddesign of organisation charqacterised by a hierarchy of authority, formalised rules and regulations that serve to guide the coordinated functioning of an organization.

Basic Postulates of the Classical Approach by Max Weber

1. Management of an organization is considered as a chain of inter-related functions. The study of the scope and features of these functions, the sequence through which these are performed and their inter-relationship leads one to draw principles of management suitable for universal application
2. Learning principles of management is done through the past experiences of actual practicing managers
3. As business environment consists of uniform cycles exhibiting an underlying unity of realities, functions and principles of management derived through process of empirical reasoning are suitable for universal application
4. Emerging new managers through formal education and case study can develop skill and competency in management concepts and practices
5. The clasasical approach also recognised the importance of economic efficiency and formal organizational structure as guiding pillars of management effectigveness.
6. Business activity is based on economic benefit. Organizations should therefore control economic incentives

Neoclassical theory of management

There are 3 neoclassical theories:

Human Relations theory : Explains the modern advancement of Human Relations Management theory which takes into account human factors like the employer-employee relationship. Human relations theory is largely seen to have been born as a result of the Hawthorne experiments which Elton Mayo conducted at the Western Electrical Company.

The important strand in the development of modern management was the increase in attention to the human factors, which has become known as the 'human relations school of management.' The core aspect of Human Relations Theory is that, when workers were being observed and included in the research, they felt more important and valued by the company. As a result, their productivity levels went up significantly. This represented a significant departure from many of the classical theories, particularly Fordism, as it went against the notion that management needed to control workers, and remove their autonomy at every step. Instead, it showed that by engaging with workers and considering their requirements and needs, company's could benefit from increased productivity.

Behavioral theory : The behavioral management theory is often called the human relations movement because it addresses the human dimension of work. Behavioral theorists believed that a better understanding of human behavior at work, such as motivation, conflict, expectations, and group dynamics, improved productivity.

The theorists who contributed to this school viewed employees as individuals, resources, and assets to be developed and worked with — not as machines, as in the past. Several individuals and experiments contributed to this theory.

Social systems theory.: Developed by Niklas Luhmann is an option for the theoretical foundation of Human Resource Management (HRM). After clarifying the advantages of using a grand (social) theory as the basic theoretical perspective, the roots of this social systems theory - the deterministic view of systems as machines, the open systems approach and non-linear systems theory - are addressed. Based on the view of social systems as autopoietically closed systems, five major contributions to a theoretical foundation of HRM are identified: (1) the conceptualisation of organising and managing human resources as social processes, thus overcoming an individualistic angle; (2) the new importance of individuals as essential element in the system's environment; (3) the abstention form far reaching or highly unrealistic assumptions about the 'nature' of human beings; (4) the interaction between various levels and units of analysis built into the theory which is essential for comprehensive and in-depth analyses of HR phenomena and (5) the openness for additional theories for which social systems theory provides the overall framework. George Elton Mayo was in charge of certain experiments on human behavior carried out at the Hawthorne Works of the Western Electric company in Chicago between 1924 and 1927. His research findings have contributed to organizational development in terms of human relations and motivation theory.

Elton Mayo's contributions came as part of the Hawthorne studies, a series of experiments that rigorously applied classical management theory only to reveal its shortcomings. The Hawthorne experiments consisted of two studies conducted at the Hawthorne Works of the Western Electric Company in Chicago from 1924 to 1932. The first study was conducted by a group of engineers seeking to determine the relationship of lighting levels to worker productivity. Surprisingly enough, they discovered that worker productivity increased as the lighting levels decreased — that is, until the employees were unable to see what they were doing, after which performance naturally declined.A few years later, a second group of experiments began. Harvard researchers Mayo and F. J. Roethlisberger supervised a group of five women in a bank wiring room. They gave the women special privileges, such as the right to leave their workstations without permission, take rest periods, enjoy free lunches, and have variations in pay levels and workdays. This experiment also resulted in significantly increased rates of productivity.

In this case, Mayo and Roethlisberger concluded that the increase in productivity resulted from the supervisory arrangement rather than the changes in lighting or other associated worker benefits. Because the experimenters became the primary supervisors of the employees, the intense interest they displayed for the workers was the basis for the increased motivation and resulting productivity. Essentially, the experimenters became a part of the study and influenced its outcome. This is the origin of the term Hawthorne effect, which describes the special attention researchers give to a study's subjects and the impact that attention has on the study's findings.

RESEARCH POLICY AND RESEARCH MANAGEMENT

There has been a clear increase in the role of the private sector in agricultural R&D, both profit and non-profit. This is due to two forces. On the demand side, it has resulted from democratization and decentralization of governance, increasing organization of civil society, and structural adjustment policies which have reduced the role and the budgetary capacity of the public sector. This has led to the privatization of many agricultural services including variety development, seeds, animal health services, marketing boards, water delivery, and technology transfer. On the supply side, advances in biotechnology, stronger intellectual property protection, improved information systems that facilitate access to scientific and market information, and more open market policies have expanded the range of technologies whose benefits can be appropriated by the private sector. For technologies which remain public goods, the supply side has also frequently been privatized under funding by foundations or national and international grants. Transfer of services to the private sector has been unevenly successful and the distribution of benefits from the change has been unequally distributed across the farm population, with some sectors benefiting while others have been left unattended. The CG focus on research management consequently needs to be broadened to understand the whole spectrum of public, non-profit, and private actors involved. Steps have been taken in this direction, with ISNAR working with NGOs, and looking at the complementary roles of the private and public sector, particularly in Latin America.

The panel consequently recommends that the CG mandate on research policy and research management be broadened to capture not just the traditional public management dimensions of R&D, irrigation, and forestry but also the public-private interface, the co-production of public goods by public and private sectors, and the role and management problems of NGOs and other non-profit organizations of civil society.

Policy and Management Decisions

We observed that most CG policy research had addressed the question of policy impact and policy advice, while little had been done on policy making and policy-making advice. Yet, the full payoff from policy impact analysis requires that it be imbedded within policy making analysis. This is because the political feasibility of policy recommendations requires assessing policy making.

For instance, recommendations of decentralized governance for participatory rural development may be vacuous, as shown by the experience with community development in India, without an understanding of the decision-making process at the local level. The policy-making game is fundamental for both positive analysis and for policy-making advice. Equally, important is to assess the credibility of policy in the sense of time consistency. If policy making institutions are not credible, for instance a Central Bank which is not autonomous from government and need to assume deficits created by soft budgets constraints, neither is monetary policy credible.

The need for political economy analysis also applies to management research. It is important in water management since much of the planned allocation of water is perverted by rent seeking. Patterns of rent seeking need to be understood to improve the efficiency of existing systems, to facilitate the transition to water markets, and to design effective water users' associations. It is also important in research management since rent seeking pressures distort the socially optimal or CG-optimal allocation of resources. In addition, not all resources that enter a research system contribute to research, with extensive deviations of funds in response to rent seeking forces in some countries.

This Panel is informed of previous recommendations made by internal and external reviewers as well as of past attempts to develop this important area of research and of modest ongoing initiatives. It recommends that every effort be made to continue to explore possibilities of making headways in research on the political economy of policy and management decisions.

Research on Institutions

An area of economic theory where there have been major advances in the last fifteen years is the "new" institutional economics. These theoretical advances have allowed to bring behaviour into traditional institutional economics and to thus understand the determinants of the choice of institutions for efficiency purposes.

This body of theory postulates that technological and institutional changes are brought about to reduce transactions and transformation costs (North and Wallis, 1994). In particular, market failures due to transactions costs or to problems of imperfect information induce the quest for institutional alternatives that allow to achieve the objectives sought (production, consumption, or exchanges) more efficiently than through markets. This theory has been highly successful in the field of agrarian institutions, for instance to explain the choice of land tenure contracts (sharecropping versus fixed rent), optimum farm size (family versus commercial farms), labor relations (permanent versus seasonal labor; time versus piece rate wages), vertical integration, the Grafting of informal financial institutions (group lending, savings and loan cooperatives), the organization of cooperative users associations (water users' associations, service cooperatives, social forestry schemes), etc. It has also been important in the field of the environment, for instance for the definition of rules for the management of common property resources and the definition of incentive contracts for the conservation of resources. Finally, it has also been highly successful in the field of industrial organization and management to answer such questions as why firms are organized the way they are, how to design incentive compatible regulatory agencies, and how to design incentive systems in firms and organizations.

In general, use of this approach has been seriously lagging in CG research. Notable exceptions include work on intra-household food distribution and inequality (e.g., Haddad and Kanbur, 1990), and work on mutual insurance schemes and consumption smoothing using the ICRISAT panel data (but largely done by non-CG scientists). We are aware of the fact that some of the research on institutional economics has been highly academic, with little of relevance for normative policy purposes and risks of engaging in marginal issues for the CGIAR. Yet, this is not the rule. Very profitable uses have been made of this approach, ranging from socioeconomic research on household behaviour under imperfect markets, to the Grafting of agrarian institutions in response to the vacuum left by state contraction, and to the design of research and management institutions for technology and water among others. In management research, institutional issues arise in the design of the optimum interface between private and public roles. For germplasm conservation, questions of incentive structures for in-situ conservation require the design of implementing and regulatory agencies.

This Panel recommends that all Centres make a serious efforts to get better acquainted with the potential that the "new" institutional economics offers for their socioeconomic, management, and policy research. It suggests that scientists concerned in different centres collaborate in systematically exploring areas where worthwhile advances could be made using this approach.

Common Property Resources

The theme of common property resources appears in the research agenda of many Centres. Indeed, this is an important issue for many resources with which the System is dealing: open access and community crop and range lands, irrigation systems, community forestry, and open access fisheries. The topic has become very fashionable in universities, in part because it is quite interesting from a theoretical standpoint. However, there is little good empirical work. Like for village and household studies, the CG could make a distinct contribution by organizing and sustaining data collection on the impact of common property on performance. Elinor Ostrom's work has attracted attention because she was able to collaborate with strong theoreticians while keeping the theoretical exercises focused on relevant questions. The CG could pursue this example, seeking collaborations with universities where there is good theory and helping make the link to empiricism and policy on the types of questions relevant for the CG.

The current large emphasis given to research on common property resources in the System seems fully warranted and this Panel suggests that this effort should be pursued, particularly in terms of solid empirical research. There is somewhat of a contradiction between the mandate given by the CG to its Centres, namely to give priority to strategic research, i.e., to research which is producing international public goods, and policy research since policy is fundamentally an issue for which the field of application is the nation state, with all its specificities and idiosyncracies. Cross-country policy studies in which IFPRI engages are appropriate to isolate the common dimensions of policy to a number of countries and to recommend a set of generic policy adjustments. In general, IFPRI's research has shied away from close intermingling with particular country governments, except at a fairly abstract level. If generic research ends up extracting generic policy conclusions and making generic policy recommendations, much of the potential value of research for policy advice may be lost. This is because, the bulk of policy advice is highly country specific as well as specific to a particular contextual instance (conjoncture), for example in terms of the particular coalition of forces that prevails at the moment when policy advice is made. There are several solutions to this dilemma.

One is to consider that IFPRI's role is more to demonstrate how to do policy research and to suggest an array of alternative policy options than to make country specific policy recommendations. If this is the case, IFPRI must be careful that the research teaching function is indeed fulfilled and that specific countries are able to extract the country level policy recommendations that derive from the analysis. Leaving this to the initiative of the reader of a generic study is too much to expect. The second solution is to engage in collaborative research, as IFPRI does, explicitly giving to country counterparts the responsibility of extracting country-specific policy conclusions. This requires more than simple co-participation of nationals to the research effort. It also means that IFPRI must assist in the national effort at extracting specific policy conclusions. This can be done in the context of preparing materials for national policy fora, as has indeed sometimes been done.

A related issue is that many of the policy studies recommend partial policy adjustments without resetting them in the country's global budgetary and political processes that establish the opportunity cost of these recommendations. For instance, most of the research on infrastructure simply recommends that more is better than less, but without establishing the opportunity cost of additional investment in infrastructure, and without asking from where funds would be coming. Again, correcting this would require developing a comprehensive view of the country's options and the processes whereby these options are selected and implemented. It would consequently require getting more immersed in the country's policy problems and the constraints under which policy options must be established.

The implication is that more comprehensive country-level policy analyses may be needed. This would significantly recast the definition of research priorities, for instance focusing on countries engaging in significant policy reforms, say Mexico or Vietnam, and proceeding with methodologies for comprehensive assessment of the set of policy options open to them in a particular time frame.

The Panel does not claim to have the answer to these difficult questions. It recommends that IFPRI and other Centres involved in policy and management research devote some thinking to the issue of generic versus country-specific policy studies before the next EPR and that this issue be addressed by the EPR itself.

Microdata

Many CG Centres and NARSs have conducted household-farm or farm surveys, for instance ICRISAT for household-farm and IRRI for farm surveys.

These data sets vary in quality and in comparability. Low-quality data are essentially useless. Sometimes, low quality is due to limited design and scope. For example, the T&V Extension Systems all collect Monitoring and Evaluation data. To this Panel's knowledge, none of these data bases has led to published work, even though millions of dollars have been spent in Africa. Sometimes, the low quality is due to poor management and control (you cannot manage these from Washington). We are not advocating that existing data be centralized. On the other hand, there clearly exists a need to better document existing data bases, both biological and socioeconomic, and to advertize opportunities for collaborative analysis, particularly for post-graduate students seeking data for their theses.

There is also a need for a more systematic process of regular surveying through the CG-NARSs system, including and especially through panel data. This is again an instance where the outposting of IFPRI senior staff for joint projects involving primary data would be mutually beneficial. There is now enough experience to proceed to a case format in panel data collection-with local variations and special studies. Agricultural research and extension systems have a stake in measuring farmers. So far, this has not been possible, but it should be possible. In general, economists are running out of data and theory has made advances which have not, for this reason, been followed by empirical validation.

POLICY AND MANAGEMENT RESEARCH IN THE SYSTEM

The overall importance of social science research in the CGIAR and the relative importance of its three components-socioeconomic, policy, and management research. These data are derived from answers to the questionnaire sent to Centres for this study. The overall allocation of CG scientists to the social sciences is 17.4% and the corresponding budget allocation is 14.8%. This indicates that the social sciences have assumed a major role in the System, raising the question, whether this share seems to be too high or not. The smaller share of budgets compared to scientists presumably reflects the fact that social scientists are cheaper because they do not require extensive laboratory facilities. Social scientists are not only engaged in research but also importantly in outreach, including institutional development, and in training activities.

In addition to IFPRI, IIMI, and ISNAR which are primarily social science oriented, some other Centres are heavily vested in the social sciences. This includes CIFOR (34.8% of its scientists), ICRAF (17.3%), ICARDA (16.4%), ILRI (14%), CIP (13.5%), ICLARM (12.5%), and CIAT (10.2%). The larger crop improvement and primarily biophysical centres such as CIMMYT (8.1%), IRRI

(8.3%), and ICRISAT (8.9%) have lower shares of their scientific staff in the social sciences. In the "Other Centres", socioeconomics is vastly more important than policy and management research: overall the distribution of scientists is 8.3% to socioeconomics, 3.2% to policy, and 0.4% to management. In the three social science centres, policy is the dominant activity due to IFPRI's weight (45.3% of all scientists), followed by management (9.8%), and socioeconomics (9.2%). These three social science centres account for 103 of the System's 937 scientists, or 12.4% of the total.

Policy Research

The lead Centre for policy research is IFPRI with a budget of US$9.3 million in core funds and US$4.5 million in complementary funds, or a total budget of US$13.8 million in 1994. At that date, IFPRI had 32 research fellows (including the DG, division directors, and a visiting research fellow) and a total of 30 social scientists. Research is organized in four divisions: environment and production technology, markets and structural studies, food consumption and nutrition, and trade and macroeconomics. Institution strengthening is managed by an outreach division. The MTP focuses on four areas of research: accelerated growth and transformation of agriculture, natural resource management policies, market economic reforms and trade policies, and household food security and nutrition. It is implementing this programme through 17 multicountry programmes that focus on strategic research issues with expected large payoffs. This research is wide ranging, covering subjects from macro problems of structural adjustment and its impact on agriculture to intra-household and nutritional effects of credit programmes and technological change policies. IFPRI is a highly visible institution, with the 20/20 Vision for Food, Agriculture, and the Environment programme as a case in point. It has generated an intense flow of publications and manages a large number of collaborative research programmes with foreign institutions and with other CG Centres. In general, the relevance and intensity of IFPRI's contributions have been highly praised in the economic profession and international community. Its scientific credibility is well established and it is widely known in LDC policy circles. According to medium-term plans and answers to the questionnaire, other Centres with involvement in policy research are as follows:

- CIAT has a small commitment to policy research, identifying only two areas of work: the dynamics of land use, in collaboration with IFPRI, and the policy determinants of sustainability.
- CIFOR has a significant commitment to policy research, with 44% of its social scientists devoted to this activity and 6.5% of the Centre's total research budget. However, it stresses the tendency of its scientists to

work on socioeconomic and policy issues in an interrelated fashion, thus making separation between the two difficult. Themes addressed include the determinants of diffusion of technical improvements in forestry, the policy determinants of sustainability of forests resources, policy options to increase employment and income from forests, equity in the distribution of benefits from forestry, and extra-sectoral influences on tropical forests including the role of the policy-making process. Specific research projects include the formulation and implementation of African forest policies, international comparisons of the intersectoral linkages with tropical forests, links between policies and livelihoods in the Amazon forest margins, and reforestation policies on degraded lands in Asia.

- CIMMYT does not have projects with a main focus on policy research, but much of its socioeconomic research has policy components. Examples of sub-projects with a policy content are analysis of the determinants of adoption of resource-conserving cropping practices in Mexico and Central America (with IFPRI and CIAT) and preparation of materials for workshops on policy reforms in Central America.
- CIP similarly does not consider that it has distinct policy, management, and socioeconomic projects, but that these activities are all integral parts of larger, umbrella-like multi-disciplinary projects, with differential weights given to the three social science components according to the nature of each project. Projects with an important policy component include the characterization of constraints and opportunities for potato and sweetpotato production in the Andes, evaluation of the impact and sustainability of potato production technologies, determinants of the expanding utilization of potato in less developed countries (with IFPRI), and post-harvest management of Andean food commodities.
- ICARDA has a large share of its scientists devoted to the social sciences (16.4%) although only 2.7% are engaged in policy research. However, boundaries between socioeconomic, management, and policy research are again stated not to be clear. Projects with a significant policy component include analysis of land tenure and public management of common property resources including rangeland and groundwater resources, management of natural resources at the farm level with emphasis on the behavioural implications of distortive policy incentives, and constraints on adoption of improved practices, particularly for range management, including the role of inappropriate incentives that lead to misuse of resources. It collaborates with IFPRI in a regional programme for integrated crop/livestock production in WANA.

- ICLARM has significant socioeconomic research but very limited explicit policy activity. Research with policy content includes the development of systems for the co-management of fisheries, and integrated resource management in small-scale mixed-enterprise farming systems. It plans to collaborate with IFPRI on themes of common property resources and fisheries policy.
- ICRAF allocates 15% of its budget and 17.3% of its scientists to social science research. It has an explicit policy research programme where the main themes are the identification and analysis of policy constraints to adoption of agroforestry technologies and the development of alternative policies to facilitate adoption.
- ICRISAT strongly stresses the need for and existence of a continuum between socioeconomic and policy research. It has had a significant programme in the socioeconomic analysis of farming systems, household behaviour, and village economies, with important policy implications. It collaborates with IFPRI on resource management research.
- IITA has a small social science programme but 41% of the social science budget is allocated to policy research. Policy research is contained in projects addressing the determinants of adoption of improved practices and systems in moist savanna agroecosystems (in collaboration with ILRI), and opportunities for improving welfare while arresting resource degradation in humid forest zone development (in collaboration with ICRAF).
- ILRI (formerly ILCA) has been actively involved in policy research with 8 senior social scientists and 4 associate social scientists, and 15% of its budget allocated to socioeconomic (12%) and policy (3%) research. However, socioeconomic and policy research are again difficult to separate. ILRI's policy programme focuses on four research themes: market and consumption of livestock products, animal health and disease control policies, macroeconomic reforms and livestock sector development, and resource management policy. It has defined four areas of collaboration with IFPRI: production systems in Sub-Saharan Africa; macro-economic reforms, regional integration, and the livestock sector in West Africa; reforms for animal-health input markets; and determinants of competitiveness of and demand for domestic dairy products in Sub-Saharan Africa.
- While mainly devoted to management research, IIMI also engages in policy research with 15.4% of its social science research budget allocated to this activity. Projects with a dominant policy component include the

strengthening of irrigation management in Egypt and improvement of the national water management programme in India.

- IPGRI has a small social science research budget (6%), but 35% of it is devoted to policy research. Themes with policy content include the analysis of policy issues relating to genetic resource management, national and international structures, and intellectual property rights issues relating to plant genetic resources and their conservation and access; and feasibility study on options for a multilateral system for plant genetic resources conservation and use.
- IRRI devotes 7.2% of its budget to social science research but only 1% to policy research. The latter includes long-term projections and policy implications of rice supply and demand in Asia (with IFPRI); and analysis of the roles of technology, price, and non-price factors in the development of rice economies of Asia.
- WARDA allocates 8% of its budget to social science research, half of it for policy analysis. Policy research includes the analysis of comparative advantages of rice production systems in West Africa and the analysis of pesticide use and policies in Côte d'Ivoire.

While principally dedicated to management research, ISNAR and IIMI are also contributing to policy research. A full characterization of policy research in the System would need to identify these contributions as well.

Management Research

Centres specialized in management research are ISNAR for research on research policy and on the management of research, and IIMI for research on water management.

- ISNAR tackles several management issues in its programmes. Some of these are not "management" in a narrow sense, such as research programme design, which is concerned with both macro-level priority setting that belongs to policy, and with institute level scientific programme management. Other aspects that have received considerable attention are monitoring and evaluation systems generally, as well as some specific thematic issues, such as organizing for natural resource management research. ISNAR's programme on "Management of Organizations and Resources" gave considerable attention to the development and installation in several countries of a NARO management information system (INFORM). Other aspects of ISNAR's research include the management of human (including gender), financial, physical, and information resources by NARSs.
- IIMI engages in a wide range of aspects of water management research that includes assessment and improvement of the performance of

irrigated agriculture, sector-level management of irrigated agriculture, improvement of public irrigation organizations, local management of irrigation systems, and the sustainable management of water delivery and disposal. Cross-cutting themes include gender issues in the performance of irrigated agriculture, health and environmental problems related to irrigation, analysis of technological options, and development of a global data base on irrigated agriculture. An important feature of IIMI's management work it that, following extensive privatizations and devolution of management to users, it has expanded its research beyond the public management of irrigation systems to the role of water users associations and the determinants of cooperative behaviour in the management of common-property water resources.

- Among other Centres, most declare not to have a specific management research component, even though management aspects may enter indirectly through socioeconomic and policy research. Centres with no explicit management research are CIAT, CIMMYT, ICLARM, ICRAF, ICRISAT, IITA, ILRI, and WARDA.
- CIFOR devotes a small fraction (1.3%) of its budget to management research. This research includes a survey on research manpower in the South African development community, and identification of the policy and management barriers to application of forestry research results.
- CIP has minimal management research that mainly enters through projects with socioeconomic or policy objectives. This is for instance the case with research on characterization of constraints and opportunities for potato production in the Andes.
- ICARDA is conducting institutional research on property rights and public/local institutions that regulate use of rangelands (in cooperation with IFPRI). It is seeking support to expand its research on the public management of underground water for irrigation.
- IPGRI does research on the design, organization, and management of plant-genetic resource activities in NARSs.
- IRRI conducts management research on the setting of priorities in rice research. This includes analysis of yield gaps, identification of research priorities and estimation of payoffs from research, and assessment of rice-research capacity in the Asian NARSs.

While principally engaged in policy research, IFPRI also conducts management research. This includes public management research which is an important component of public policy research. In addition, IFPRI has become increasingly involved in research on research management, water management, and forestry management, generally in collaboration with the corresponding specialized Centres, namely ISNAR, IIMI, and CIFOR.

Socioeconomic Research

This is the most important area of social science research in the Centres other than IFPRI (policy research) and ISNAR and IIMI (management research). The nature of this research has been analysed in another TAC special study and we only make general reference to this type of research principally because it is both essential for policy and management research and often hardly separable from these two. Even IFPRI, ISNAR, and IIMI do considerable socioeconomic research since analysis of the behaviour of consumers, producers, households, communities, associations, managers, rent seekers, and policy makers are fundamental cornerstones of policy and management research.

In principle, socioeconomic research conducted by Centres where biophysical research is dominant has as a purpose to guide research resource allocation toward priority areas, help in technology design, assess the determinants of adoption of new technologies, and determine the impact of diffusion of these technologies on welfare and the environment. For this type of research, the biophysical centres have a comparative advantage over IFPRI as they can most easily mobilize the necessary interdisciplinarity with biological scientists. This research has an important service function in giving guidelines to researchers and the Centres. In fact, socioeconomic research covers a wide array of issues, showing that it has become an integral component of research programmes as research questions and the CO mandate have become increasingly more complex and holistic. No longer is socioeconomic research introduced as a prop to the definition and diffusion of technological change, but it has become an undissociable component of the search for solutions to complex interdisciplinary problems. This is, of course, the main reason why social science research has acquired growing importance in the CG's total research portfolio.

It is worth noticing that 17 out of 78 scientists devoted to socioeconomic research in the CGIAR, or 22%, are at ICRAF. This percentage raises to 25% when IFPRI, ISNAR, and IIMI are excluded. The Panel could not look into the reasons for this disproportionately high presence of socioeconomic research at this particular Center.

Some examples of the wide array of issues addressed by Centres' socioeconomic research is indicated by the following answers to the Centres' questionnaire:

- CIAT focuses mainly on the analysis of prototype production systems and participatory research methods.
- CIFOR conducts research on household strategies and community differentiation in the production of non-traditional forest products in

Sri Lanka, household economics and forest use in Brazil, interactions between environment and population in Thailand, forests and food security, income generation and incentives for forest management, and the social dimensions of sustainability criteria.

- CIMMYT's socioeconomic research addresses questions of global maize research and the maize seed industries, the conservation of germplasm diversity, impact assessment of research in wheat and maize, priority setting and research resource allocation (with IFPRI), the determinants of adoption of long-term conservation investments, and the diagnosis of natural resource management issues.
- CIP's socioeconomic research is integrated in the same topics that are part of its policy research programme mentioned above.
- ICARDA's socioeconomic research focuses on the economic analysis of production systems and the exante and expost analysis of new technology options. Projects address integrated crop/livestock production, water resource management, water harvesting, and farmers' participation in and use of local knowledge at breeding barley for specific adaptation.
- ICLARM conducts socioeconomic research on the dissemination and evaluation of genetically improved fish strains, the bioeconomic modelling of capture fisheries, and bioeconomic models of coastal aquaculture systems.
- ICRAF's socioeconomic research includes the development of typologies of land-use systems and farmers, the development of ecological-economic models which predict the adoption potential and sustainability of improved agroforestry practices, and the evaluation of the actual ecological, social, cultural, and economic impacts of adoption of agroforestry practices.
- IITA's socioeconomic research includes improving postharvest systems, the diversification of humid forest production systems, improvement of yam-based systems, and the determinants of cassava productivity under various ecological conditions.
- ILRI's research includes a wide range of themes focusing on the economic and environmental impacts of livestock disease control, the economics of mixed crop-livestock systems, the socioeconomic analysis of livestock production and natural resource management, the economics of peri-urban dairy systems, and impact assessment and priority setting studies.
- IPGRI's research addresses the role of gender in plant genetic resource conservation and use, documentation of indigenous knowledge for farmers and scientists, and the effect of conservation and use of

germplasm on socioeconomic development through agricultural development.

- IRRI's socioeconomic research focuses on the determinants of adoption of knowledge intensive rice technology, understanding the risks in rainfed rice farming, and the impact of rice technology on rural income distribution and poverty.
- WARDA's research agenda includes farmers' perceptions and adoption of new rice varieties, the economics of rice-based cropping systems, agricultural intensification and women, economic constraints to rice based rotations in the Sahel, differentials in economic efficiency for small and large rice farmers, and the rate of return to rice breeding research.

THE MANAGEMENT OF RESEARCH AND DEVELOPMENT

Ideas concerning the management of research and development have undergone rapid development in the last few years and several new management methods have come into use in companies. For eighty years or more R&D activity expanded in medium and large companies, but it has perhaps now reached saturation level. Scientists and engineers in companies often pursued research of their own devising while having only slight awareness of the commercial requirements.

At the same time corporate managers tended to leave R&D to its own devices, both from a feeling that this was something they did not understand, and because R&D issues are never so urgent as the problems of production start-ups and market launches. This was often unwise because the products being launched determined the futures of the companies, yet were sometimes selected for development by R&D without due attention by senior management. Company spending on R&D varies from 0.1 percent of turnover to over twenty percent, with an average of perhaps four percent Despite its importance R&D has received scant attention in management teaching and textbooks. R&D management almost never features in business school syllabuses or faculty lists, or in the indexes of standard management texts. It is however covered in really excellent handbooks of management.

New Thinking

Pressure for change has come from the deep worldwide recession of the early nineties causing companies to look closely at all classes of expenditure, and also from rather belated recognition of technology and innovation as competitive elements. It has been recognised that although competition does indeed take place on the classic grounds of efficiency, price, promotion and marketing, ownership of a technology can also give a profound advantage. Mastery of the difficult art of making silicon chips restricts that business to a few, while legal ownership of drug formulations enables pharmaceutical companies to make monopoly profits. Both of these advantages originate in the R&D laboratories of the companies concerned. Innovation is of course broader than technology development by R&D, and can occur for example when business advantage is gained simply by buying-in new technology machinery from outside. Innovation is a concern of all the company, and the role of R&D in this is best defined simply as that part of innovation which happens to be done by the people in the R&D department.

Major contributions to new thinking about the management of R&D are the concepts of "third generation R&D" brought together by the consultancy company Arthur D Little , the Stage-Gate system of managing new product development , portfolio management theory of R&D projects and the methodology of technology foresight. These provide difficult but convincing tools for managing an activity that has caused much anguish in the past, and they enable constructive dialogue to take place between R&D and the rest of the company.

The Nature of R&D

R&D work is intrinsically risky. Almost by definition R&D tries out new ideas to see how well they work, seeks information that is as yet unknown, and hopes for ideas as yet unthought of. This means that success cannot be commanded, and most R&D is managed in terms of probabilities. The laboratory is gambler territory. It is useful to contrast R&D with design, where the assumption is that all the knowledge required exists, and if the design is competent the product will work. Products that are designed tend to be of modest profitability because the knowledge required is likely to be fairly widely available. By moving into the unknown, R&D accepts risk in return for the chance of greater reward. Managers of R&D have to be able to live with a lot of failure.

MARKETING TECHNIQUES

Marketing is the process of performing market research, selling products and/or services to customers and promoting them via advertising to further enhance sales. It generates the strategy that underlies sales techniques, business communication, and business developments. It is an integrated process through which companies build strong customer relationships and create value for their customers and for themselves.

Marketing is used to identify the customer, to satisfy the customer, and to keep the customer. With the customer as the focus of its activities, it can be concluded that *marketing management* is one of the major components of business management. Marketing evolved to meet the stasis in developing new markets caused by mature markets and overcapacities in the last 2-3 centuries. The adoption of marketing strategies requires businesses to shift their focus from production to the perceived needs and wants of their customers as the means of staying profitable.

The term *marketing concept* holds that achieving organizational goals depends on knowing the needs and wants of target markets and delivering the desired satisfactions. It proposes that in order to satisfy its organizational objectives, an organization should anticipate the needs and wants of consumers and satisfy these more effectively than competitors.

8

Research in Quantitative and Qualitative Measurement

INTRODUCTION

Quantitative research relies primarily on numbers as the main unit of analysis. It is more commonly used as a primary method in scientific and clinical research, such as drug trials or laboratory experiments where tests may need to be repeated many times, for example to ensure that a new drug is safe. Although quantitative methods, such as surveys, are used in educational research, the vast majority of research is relatively small scale, intensive, focused on change and involves human perceptions. Educational research relies much more heavily on qualitative methods. One of the most common instruments to gather numerical data in education (particularly in evaluation of programmes) is the questionnaire survey, using a series of closed questions to which responses are given against a Likert or other type of scale.

Open questions can also be included to gather richer data. Large amounts of data can be gathered from a wide number of people and the results can be analysed by computer (either by an optical mark reader or through an online survey instrument such as 'Survey Monkey'), thus making it fairly straightforward to research a large sample of respondents. Survey questionnaires can be given out and collected face to face, sent by post or posted online. If achieving a high response rate is important, then note that the less personal involvement there is with potential respondents, the lower the response rate. So, typically, online surveys may have a response rate of under 20 per cent, whereas if the questionnaires are given out and collected face to face, you may achieve a very high response rate.

Qualitative research

Qualitative research relies primarily on words as its unit of analysis and its means of understanding. However, it can also use voice tone, loudness, cries, sighs, laughs, and many other ways of human communication. The words may be spoken in individual interviews (face to face or on the telephone) or groups, or they may be written, so you may have to analyse the spoken words of an interview, focus group or conversations (for example between a patient and health worker), or the written words of an account or description or diary record.

On the whole, qualitative research tends to be small scale, simply because it is hugely labour intensive. For example, interviews or focus groups will usually need to be transcribed before they can be analysed. In addition, the researcher is often more involved with the person producing the words, and so it is sometimes helpful for others to conduct the analysis; again this can be costly.

Having said that, nothing else can provide the same level of richness as qualitative data, and at the very least, adding space for respondents to provide some words to describe what might be otherwise gathered by numbers is immensely useful to the researcher, and may even, in some situations, be a help to the subject.

Qualitative methods range from the classification of themes and interconnections, content analysis, grounded theory and discourse analysis, and reliability and validity are just as important as they are in quantitative analyses. There are computer programmes to assist in analysis, and although these might not necessarily save time, they often offer more systematic ways of coding data and identifying connections and themes.

Number of Research Methods

Triangulation is the term used for bringing a number of research methods to bear upon a question. For example, to study the effect of threatened closure of a hospital on its staff you might want to interview a selected few at different levels and in different professions; a questionnaire survey of almost everyone; personnel data of sickness, absence, turnover, etc.; a network analysis of rumour; an ethnographic study of a long-stay ward looking at staff and patients interactions; and so on. A triangulation of methods such as these would provide an exceptional in-depth look at such an event, but it is also useful simply to combine some form of qualitative and quantitative data, as suggested above.

TYPES OF QUESTIONS AND CORRESPONDING METHODS

The insight that the method should follow from the question, now pretty obvious and something graduate students often hear from their professors, led to another conundrum. There are so many research questions, how do we say anything cogent about method? The Committee reasoned that most scientific research questions were of three general types: (1) What's happening? (2) Is there a systematic (causal) effect? And (3) what is the causal mechanism or how does it work?

What's Happening?

The question—what's happening?—asks for a description. We could ask this question in a materials science laboratory or in a middle school. We could describe the years of experience possessed by elementary school teachers in the U.S., or the types of science instruction students receive, or the changes in students' mathematics achievement over 20 years. In these cases, statistical estimates of population parameters could be obtained from available national surveys. Or we could describe what school, home and community look like though the eyes of an inner-city youth using ethnographic methods. Or we could describe different approaches to the assessment of learning in colleges and universities that have been nominated as "exemplary" using case study methods.

Holland and Eisenhart's (1990) study exemplifies scientific research into what's happening, and, as we shall see, beyond. They were concerned about explanations for why so few women pursued careers in non-traditional majors such as science: women were not well prepared before coming to college, women were discriminated against in college, women did not want to compete with men for jobs. They began their study by *describing, in depth over several years* through *ethnography*, the lives of 23 volunteer women from two small public residential colleges.

Half of these women pursued traditional careers and the other half non-traditional careers. They were matched on grades, college majors, college activities, and college peers. Based on extensive, detailed data collected through participant observation and interviews, Holland and Eisenhart found that what best described these women's academic pursuits were, *contrary to popular conjecture,* how they viewed the value of schoolwork, what their reasons were for doing school work, and how they perceived financial and opportunity costs.

Simply put, detailed description of the college lives of these women portrayed their career trajectories in a very different light than the distal conjectures based on statistical data.

Now if you are thinking, "well, this is just idiosyncratic description and is suspect," Holland and Eisenhart one up you. They took the next step, going beyond description and entering the arena of model testing by predicting with their descriptive models what each of the 23 women would do after college: continue in school, get a job in her field, get a job outside her field, get married, etc. At the end of 4 years and another 3 years, they followed these women up with telephone interviews. In *all* 23 cases, predictions based on their model of commitment to schoolwork were confirmed. In all cases, their model provided a better predictor than did data on precollege preparation (grades and courses taken), discrimination against women, or feelings about competing with men.

Is There a Systematic (Causal) Effect?

Questions about effects are, ultimately, questions about causal effects. Did *x* cause y? Perhaps the most widely known study of systematic effects in education is the Tennessee randomized trial on class size reduction. The question posed by the Tennessee legislature was whether reduced class size would improve students' achievement. To this end, within 79 schools across the state, a total of 11,600 students were randomly assigned to a regular class (22-26 students), a class with an aide (to decrease the student/adult ratio economically) or to reduced class size (under 13-17 students). Three findings stand out: (1) students in small classes outperformed students in the other classes, (2) minority students particularly benefited, and (3) the effect persisted when students returned to regular class sizes. Here the method, randomized trial, followed directly from the legislature's question and was feasible and ethical to implement. In such cases as this, randomized trials are the preferred method for ferreting out causal effects.

There are many cases, however, when randomization may not be feasible. Such cases include the effect of smoking on health and longevity, and the effects of hunger, alcohol use, drug use or child abuse on students' academic performance. For these research questions, other methods need to be used, are available, and include quasi-experiments (control and experimental groups without random assignment), correlational studies using large-scale probability-sampled data sets (that adjust for selectivity bias), and various time series designs. To be sure, as you move away from the randomization in some cases uncertainties increase; nevertheless, causal interpretations are possible and replication is important to increase confidence in the interpretations.

Application of Correlational

Loeb and Page's (2000) study of teacher salaries exemplifies the application of correlational (structural) modeling in a situation where random assignment is unlikely. They asked, "If teacher quality affects student achievement, why do studies that predict student outcomes from teacher wages produce weak results?" That is, shouldn't teachers' salaries reflect, at least to some degree, teacher quality after controlling for other things?

Loeb and Page tested two competing models. One was the usual production function model that links inputs (salary) to student outcomes (dropouts in this case) after controlling for relevant variables.

The second model followed their reasoning that there are other things in the lives of teachers than salary that may have meaning, and there also may be local job markets that provide attractive alternatives to teaching in the area. So their second, competing model incorporated opportunity costs into the production function: non-pecuniary rewards and competition in the local job market. They replicated prior research with the usual production-function model, showing a weak effect of salaries on outcomes. However, once they adjusted this model for opportunity costs (non-pecuniary and job market incentives), they found that raising wages by 10 per cent reduced high school dropout rates by 3-4 per cent.

Three points about studying causal effects seem appropriate here. First, in dealing with causal assertions we are always trying to rule out all the possible counter hypotheses that we know of at the time. As a research programme moves along, new challenges (counter hypotheses) arise and get ruled out; in this way confidence increases in the causal interpretation. Oftentimes we don't know all the counter hypotheses; challenges arise with novel counter-interpretations, and research and debate continues as it has with the Tennessee study. This type of debate—hypothesis/counter-hypothesis—is the basis of science and should be looked upon positively and not as "backbiting" among scholars with different views when the issue is one of interpretation; it is backbiting when personal attacks are made.

A second point has to do with the role of description in causal studies—what's happening? When feasible, descriptive research should be used in causal studies to help us understand, as fully as possible, what "treatments" were actually implemented, and to reveal what possible causal mechanisms might be operating.

And the third point is that establishing a causal effect may be necessary when possible but not sufficient in policy and practice. The questions of mechanism and context inevitably should arise in order to design education policies or practices. We need to understand how interventions were articulated and implemented in diverse contexts with whom, under what conditions with what resources in order to design more than superficial education policy.

Causal Mechanism

The third type of research question focuses on the mechanism that creates a causal effect. For example, reducing class size seems to have a salutary effect according to the Tennessee study. But what was the mechanism that caused the effect and why did it persist even after students returned to regular class sizes Was the effect due to an increase in the number and personal nature of teacher-student contacts or to less off-task student behaviour or to the level of student engagement?

Empirical Studies of Mechanism

Empirical studies of mechanism, following on studies that have established causal effects are most common. Bryk, Lee and Holland (1993) sought to understand the causal mechanism(s) underlying the causal evidence that Catholic schools outperform public schools in the U.S. This longitudinal study used both qualitative (*e.g.*, case studies of effective Catholic schools) and quantitative data to address the mechanism question. Three potentially explanatory models were tested: (1) sector effects only (spiritual and private characteristics of Catholic schools), (2) compositional effects (kinds of students attending Catholic schools), and (3) school effects (school operations contributing to school context). A combination of models, characterizing "...the *coherence* of school life in Catholic schools ... most clearly accounts for it relative success in this area".

Nevertheless, there is another way to approach the question of mechanism—namely, to build an artifact based on a causal theory and establish its causal effect. Studies such as "design experiments" or "design studies" posit a theory with a causal mechanism, and design educational artifacts (*e.g.*, a curriculum, a computer application) and iteratively test them out in complex real-world classroom environments, revising both artifact and theory along the way. Once evidence accumulates to suggest a causal mechanism, the onus, of course, is on design researchers to then establish generalizable causal effects.

COMBINE QUALITATIVE AND QUANTITATIVE RESEARCH

Qualitative studies in psychology can be fascinating and insightful but they may leave readers with a quantitative disposition worrying about the generality of their findings. Quantitative studies, on the other hand, whilst providing data from larger and more representative samples, seem more mechanical and arid to qualitative researchers. But both methods have advantages and disadvantages and the results from different methods can complement each other.

This paper considers these issues in the context of studying the problems that students face in essay writing. It contrasts the recent results from a qualitative study with those from a quantitative one (described in this paper). At the time of writing we have copies of four of Street and Lea's articles in front of us. Only in the major one are we told how many students, lecturers and institutions were involved in their study:

'The analysis of the research data has concentrated on the differing interpretations and understandings of academic staff and students with regard to academic writing within two contrasting university settings. 10 interviews were conducted with staff in the older university and 21 students were interviewed, either individually or in small groups. At the new university 13 members of academic staff and 26 students were interviewed.'

But these are the only numerical data given. We are not told, for example, what proportion of the students are men or women, traditional-entry or mature students, and what disciplines they are studying. And there are no quantitative data to qualify their results.

The actual results are of great interest. Street and Lea differentiate between three different, but overlapping, sources of difficulty for students writing essays and reports, and how their institutions deal with them.

These are:

- Difficulties with 'interpretation' *e.g.*, knowing what is expected within and between different departments, and even between different tutors within the same department. These are largely ignored by many tutors, but students' ignorance of these matters is criticised in their feedback.
- Difficulties with 'deficits' *e.g.*, grammar, spelling, punctuation and style. These are mainly dealt with by study skills courses, and handouts/ guides for students.
- Difficulties stemming from 'institutional failings' - *e.g.*, the inability of staff to mark written work within a reasonable time span and with sufficient detail because of limited staff resources and the large numbers of students. These difficulties are ignored, defended or deplored by different members of staff.

Street and Lea place their greatest emphasis on the second area of difficulty - one that has not really been explored in previous research on essay writing. But are they right to do so? How widespread are the difficulties faced by their 47 students, and which do they find the greatest burden? We cannot answer questions such as these because of the qualitative nature of their reports.

Nonetheless, the findings, the commentary and the suggestions are so interesting that we set out to gather some more - quantitative - evidence to try to build upon this pioneering study.

Questionnaire Method

We devised a questionnaire on essay writing that addressed the three concerns discussed by Street and Lea. Thus there were questions about students' experiences of difficulties connected with 'deficits', 'interpretation' and 'institutional failings'. The questions were largely based upon the comments and discussion provided in Street and Lea. We gave this questionnaire to 102 *second-year* psychology students attending a lecture at Keele University at the beginning of their first-semester.

These students were asked to complete the questionnaire with reference to the difficulties that they might have faced when writing their essays and reports in their *first-year* at Keele. Students at Keele typically study two principal subjects and two subsidiary subjects in their first-year, so they would have had to write essays/reports in four different subject matters during the year. (It is this kind of complexity that Street and Lea had in mind in their study and which has not been commented on before.)

In our study the responses from two overseas students who had not been at Keele in their first-year, and from six students studying conductive education (who did not follow a normal subsidiary programme) were deleted, making a total of 94 respondents. (The total number of students in this cohort was 146 - so our data come from 64 per cent of the class.)

QUANTITATIVE RESEARCH

Believing that the research question was even more fundamental than the paradigm one felt allegiance to, several years ago we began to discuss the qualitative-quantitative debate from that perspective. The dichotomy and the debate disappeared, and the ideas presented here began to develop. This book describes our stance at a point in time, not the conclusions of our ideas, which continue to emerge, to grow, and to build from our work as researchers and as teachers.

While clearly a work in progress, which continues to evolve, the framework of an interactive continuum presented here has been enlightening to colleagues and students who operate within the current world of often-misunderstood and frequently debated paradigm shifts. At the conclusion of this chapter, the reader should be able to:

- Describe the history of qualitative and quantitative research methods and the debate about their relative values
- Describe the typical purpose and outline of qualitative research
- Describe the typical purpose and outline of quantitative research
- Discuss the advantages and disadvantages of a dichotomy versus a continuum conceptualization of research design

Qualitative and quantitative researches have philosophical roots in the naturalistic and the positivistic philosophies, respectively. Virtually all qualitative researchers, regardless of their theoretical differences, reflect some sort of individual phenomenological perspective. Most quantitative research approaches, regardless of their theoretical differences, tend to emphasize that there is a common reality on which people can agree.

From a phenomenological perspective, Douglas and Geertz believe that multiple realities exist and multiple interpretations are available from different individuals that are all equally valid. Reality is a social construct.

If one functions from this perspective, how one conducts a study and what conclusions a researcher draws from a study are considerably different from those of a researcher coming from a quantitative or positivist position, which assumes a common objective reality across individuals. There are different degrees of belief in these sets of assumptions about reality among qualitative and quantitative researchers. For instance, Blumer (1980), a phenomenological researcher who emphasizes subjectivity, does not deny that there is a reality one must attend to.

The debate between qualitative and quantitative researchers is based upon the differences in assumptions about what reality is and whether or not it is measurable. The debate further rests on differences of opinion about how we can best understand what we "know," whether through objective or subjective methods.

William Firestone (1987), in an article in the "Educational Researcher", differentiates qualitative from quantitative research based on four dimensions: assumptions, purpose, approach, and research role. Regarding assumptions, Firestone asks: is objective reality sought through facts or is reality socially constructed? Related to purpose, he asks: is it looking for causes or for understanding? To determine approach, he asks whether the research is experimental/ correlational or a form of ethnography. Lastly, related to the researcher's role, he asks whether the researcher is detached or immersed in the setting.

Shaker (1990), in a discussion of programme evaluation models, presents them as a metaphorical journey—moving from quantitative perspectives in the past to more recent naturalistic and qualitative assumptions. While positing a chronological continuum, Shaker would not seem to oppose our notion of question-driven research and evaluation. While he describes the "new identity" for evaluation as being "based on naturalistic approaches," he places this in the context of a "pragmatic commitment to finding methods that yield results in practice as we find it, rather than as we wish it to be".

The qualitative, naturalistic approach is used when observing and interpreting reality with the aim of developing a theory that will explain what was experienced. The quantitative approach is used when one begins with a theory (or hypothesis) and tests for confirmation or disconfirmation of that hypothesis.

It is important here to set the stage for abandoning the dichotomy. To do so, we examine a few of the key events in the chronicle of scientific evolution that established the debate in the first place. As long as one view of how we can explain the workings of the world reigns supreme, there is no debate.

The debate rests on a dichotomy characterized by a lessening of the dominance of one paradigm over another, leveling the playing field so that the debate could occur. In fact, the debate may be but one more phase in the ebb and flow of an everchanging philosophy of knowledge. For example, in The Enlightened Eye, Eisner (1991) cautions against the dichotomy and asserts that qualitative and quantitative research can be combined. He warns against qualitative researchers merely adopting a "soft form of positivism".

The genesis of the current qualitative-quantitative debate in educational research occurred as far back as 1844, when Auguste Comte claimed that the methods of natural science could be justified in studying social science. Science, in this view, is the collection and study of facts that can be observed through sensory input. These are the traditional data investigated by natural scientists—the physicists, the chemists, the biologists.

This view holds that true science is accumulated through the study of phenomena that can be physically sensed, observed, and counted. The "unknowables," as Herbert Spencer described them in his 1910 essay, those things that cannot be sensed but might rely on reason or thought, are banished from scientific investigation. Both Comte and Spencer were positivists.

Interestingly, this "positivism" was a move away from a more speculative, more "unknowable" view. It was a move away from relying on theological and metaphysical explanations of the world. It was a move toward what could be "positively" (confirmed through sensory data) determined. The philosophy maintained a grip on social science from the late 1800s through the early 1900s.

In the early 1900s, John Dewey, among others, questioned the absolutism of this position, viewing science as not separate and distinct from problem solving. His pragmatism considered science less rigidly than did the positivists. In his Sources of a Science of Education (1929), written some time after his initial speculations, he pointed out that practice should be the ground of our inquiry. Because of the value placed on experience for learning and the emphasis on practice, he appreciated the deeper complexity of what educational and social scientists study.

During the same period, a group of scholars who made up what became known as the famous Vienna Circle met and developed a new philosophy of science, logical positivism. Supporting Comte's positivism, they combined it with the symbolic logic of mathematics. Hypotheses derived using the rigor of mathematics (the symbolic) could be combined with fact gathering (the positivism) to test their confirmability (which was eventually modified to disconfirmability).

Although counter to an impetus by Dewey to diffuse the positivistic assumptions made by researchers, this hypothetico-deductive system was dominant in the middle years of the 20th century in psychology and sociology. Education, which borrowed traditions of inquiry from these disciplines, was affected as well. The respect for precision in measurement, mathematically systematic tests of hypotheses, and a quest for value-free science solidified this paradigm.

During the 1940s and 1950s, the quantitative paradigm dominated the social science and the educational research scene. Behaviourists and organizational theorists utilized empirical fact gathering and hypothesis testing almost exclusively in studying educational and social phenomena. In the mid- 1960s, while the quantitative perspective continued to prevail, a shift began as skepticism toward the domination of logical positivism and the evident chasm between human social systems and mathematical logic grew. New epistemologies began to emerge that acknowledged, for example, the value-laden nature of human social interactions. That human beings construct reality for themselves and that knowledge itself is transmitted in social ways were beginning to be assumed. Questions arose about the tenability of applying natural science methodology to these complex human dynamics.

In 1962, in The Structure of Scientific Revolutions, the most significant work on this issue, Thomas Kuhn explored the shifts in science's dominant paradigms. His doctorate in theoretical physics led him to look back into the history of science as he sought to know more about its foundations. He describes how, by randomly exploring the literature, he was exposed to Jean Piaget and, in the late 1950s, to a historical analysis of social science and psychology. Kuhn's study of methodology drove him to leave physics and become a historian of science.

He conceptualizes the notion of paradigms, "universally recognized scientific achievements that for a time provide model problems and solutions to a community of practitioners," and proposes that competing paradigms emerge chronologically when the dominant one no longer serves the explanatory needs of the scientific community.

For the most part, using the context of physics from the perspectives of Sir Isaac Newton and Albert Einstein, Kuhn explains these periods of competition, or scientific revolutions, in the natural sciences. He acknowledges that competing paradigms can possibly coexist on equal footing following such a revolution, or "paradigm shift," although, he cautions, it may be only rarely possible. He proposes that the predominant paradigm affects researchers not only methodologically but also in how they see the world. Kuhn's conceptualization of "paradigm" has been reinterpreted by others since his work, and many definitions are incorporated in the literature of the 1990s.

The quantitative paradigm continued to reign over social science and, according to Culbertson (1988), prevailed in education until the mid-1980s. At that time the logical positivists were losing supremacy. (The strong traditional bias toward quantitative science might even be based on Americans' preference for facts we can observe and count, a sense that that's what science "is.")

Concurrent with Kuhn's early notions of paradigms in the 1960s, society was undergoing radical changes. While some began to question the efficacy of the positivists' tools in explaining human organizational and social phenomena, education was moving into a more complex social context. Culbertson points to such 1960s and 1970s issues as racial integration, poverty, equal opportunity, schools as tools in global economic competition, the Soviet Union's threat to our math and science preeminence, and the need to account for the success and failure of the nation's children and posits that, in this context of increased complexity, some began to search for policy tools that the quantitative paradigm did not seem sufficiently able to explain.

That education served economic, political, and policy ends enhanced the opportunity for scholars interested in the culture of schools to begin to use anthropological strategies in their inquiry. These same interests fed the scholars' attempts to approach their research from the perspective of the critical theorists, as well as that of the feminists. Although always an important issue, the policy makers' interest in the world of classroom practice grew, and they increasingly expressed concerns that research and practice were unconnected and that this disconnection was in part due to the use of tightly controlled laboratory-like quantitative assumptions. A move among some social scientists in the direction of deriving theory from practice, rather than the other way around, characterized this change as well.

Graduate programs preparing educational and social science researchers as well as professional journals have increasingly directed their attention toward qualitative research. Allotting time and space to what had been considered the "alternative" paradigm led to wide discussions in the journals and at professional meetings. The editors of the American Educational Research Journal, for example, announced in 1987 that particular emphasis on qualitative methodology would be forthcoming as they evaluated manuscripts.

This was a major legitimation of the paradigm for educational researchers. A plethora of books, articles, and presentations on the trustworthiness of the qualitative paradigm materialized. Some extolled the virtues of qualitative research as the only avenue to "truth," while others claimed that only by holding onto the quantitative traditions can we have confidence in our knowledge base.

In many forums the debate was manifest. Which is more scientific: the deductive methods of the logical positivists (quantitative researchers) or the inductive methods of the naturalists (qualitative researchers)? Can the results of qualitative research be generalized as are the results of quantitative research? Can science be value laden (qualitative) or only legitimate if value free (quantitative)? What epistemological assumptions are violated by adopting one paradigm or the other?

While to some the debate has ended, to others, especially those we encounter in researcher-preparation programs, the debate has either not yet materialized to the full extent of its fury or continues unabated. Our strong sense is twofold. First, we continue to prepare students for an "either-or" world, a dichotomous world that no longer exists.

We still prepare students who leave our colleges and universities with a monolithic perspective. Either they become well-trained statisticians, or they become cultural anthropologists, methodologically weak in asking research questions and in justifying either one or the other set of strategies. Second, researchers in education and in the social sciences have not yet constructed a way to ensure their success in utilizing both paradigms. The interactive continuum model in this book serves as a kind of framework directed toward both those needs.

The dichotomy of qualitative and quantitative research is one we deny but one we exploit here for heuristic purposes. The dichotomy, while not an ontological construct, does allow us to separate the idea. We slice it thin to examine it and make the case in this chapter that it does not exist in the scientific research realm. All research in education stands on basic underlying assumptions.

This is true for quantitative methods as well as qualitative methods. To the extent that these assumptions withstand the scrutiny of scientific inquiry, the methods can be supported, taught to novice researchers, and used professionally and ethically without reservation.

Since the mid-1980s when quality in all educational professions came under public review, it has become particularly crucial to delineate the foundational bases of educational research. Within the realm of this book, such bases will be examined.

Qualitative Versus Quantitative

All behavioural research is made up of a combination of qualitative and quantitative constructs. In this book, the notion of the qualitative-quantitative research continuum, as opposed to a dichotomy, is explored on scientific grounds. We believe that conceptualizing the dichotomy (using separate and distinct categories of qualitative and quantitative research) is not consistent with a coherent philosophy of science and, further, that the notion of a continuum is the only construct that fits what we know in a scientific sense. A secondary theme is equally important; that is, what are known as qualitative methods are frequently beginning points, foundational strategies, which often are followed by quantitative methodologies.

Qualitative research methods are those generally subsumed under the heading ethnography.

Other headings and names include case studies, field studies, grounded theory, document studies, naturalistic inquiry, observational studies, interview studies, and descriptive studies. Qualitative research designs in the social sciences stem from traditions in anthropology and sociology, where the philosophy emphasizes the phenomenological basis of a study, the elaborate description of the "meaning" of phenomena for the people or culture under examination.

This is referred to as the verstehen approach. Often in a qualitative design only one subject, one case, or one unit is the focus of investigation over an extended period of time. According to Glaser and Strauss (1967), qualitative data are often coded a posteriori from interpretations of those data.

Quantitative research, on the other hand, falls under the category of empirical studies, according to some, or statistical studies, according to others. These designs include the more traditional ways in which psychology and behavioural science have carried out investigations. Quantitative modes have been the dominant methods of research in social science.

Quantitative designs include experimental studies, quasi-experimental studies, pretest-postest designs, and others, where control of variables, randomization, and valid and reliable measures are required and where generalizability from the sample to the population is the aim. Data in quantitative studies are coded according to a priori operational and standardized definitions.

It is necessary to adopt some standard by which one can measure whether the qualitative, the quantitative, or a continuum that includes both methodologies is the most effective mode in reaching truth. We assume the standard of science as a way of knowing.

Mouly (1970) asserts that, although there are two ways other than science to "know" something (i.e., "experience" and "reasoning"), only through science can we generalize and provide for theory building. Some would have us believe that we can know something based on "authority" This basis has similarly been discredited because of the frequent inability to verify the facts, as well as the conflicting points of view among authorities.

Other philosophers go even further and suggest one can "know truth" also through "serendipity," "intuition," "compromise," and "consensus." Conjecture surrounding how we can know about truth, repeatable and verifiable truth, runs the gamut from "faith" to simple sensory perception. The assumption here is that science, as reflected in the scientific method, is the only defensible way of locating and verifying truth. Therefore, the criteria for comparison of the constructs underlying the dichotomy (qualitative vs. quantitative) and the interactive continuum (qualitative to quantitative to qualitative, etc.) are their scientific bases

The search for knowledge (or "truth") is the purpose of research. This search and, concomitantly, this research is most effective when built on the scientific method. In the ongoing debate between the positivists and the naturalists we tend to support the idea that the modern-day scientific method is both inductive and deductive, objective and subjective.

Design validity is more likely to be built into studies when the researcher is open to both paradigms rather than precluding one or the other. When faced with the question, "Which is better?" we would refuse to answer; indeed, we would be unable to answer, given the choices presented. There is no such answer. The better paradigm (qualitative or quantitative) is the one that serves to answer the specific research question.

We began our thinking on these issues over a decade ago. Our thoughts began to solidify in an interactive continuum model in 1985. Others have written about integrating qualitative and quantitative methods. Cook and Reichardt (1979) predates our original work and, like us, they suggest that the researcher's method can be separated from the researcher's worldview. Their book differs from ours in that their ideas are presented in an introductory essay to a collection of essays by research methodologists.

Their purpose was to bring together the combined works of many who were then struggling with the issues. Michael Patton (1980) presents a diagram of what he calls "mixed paradigms" in his book, Qualitative Evaluation Methods. His conceptualization, like ours, acknowledges that, between the qualitative and quantitative paradigms, there is a continuum of methods. His book, however, addresses qualitative methods only. It is not an exhaustive examination of assumptions, methods of research, and ways to critique research studies as we intend ours to be.

Creswell (1994), too, has authored a volume, Research Design: Qualitative and Quantitative Approaches, and he intends it to assist the researcher in making decisions about design. His book seems most closely focused on writing a dissertation proposal, and it is organized in that sequence. It does not include critiquing research as ours does, and he does not present an overall model of his thinking. The book is replete with examples from both qualitative and quantitative studies.

Our book contributes to the current discourse on research methods and assumptions underlying social science research by Depicting an overall model of qualitative-quantitative interactive continuum Suggesting ways to assess quality of published research Providing a strong emphasis on validity. In the last decade, a debate has continually raged as though one or the other paradigm should eventually win. Discounting the debate is not the issue of importance.

The key issue, we believe, should be improving the quality of research through an integrated way of viewing qualitative and quantitative research methods. Both paradigms coexist in the world of inquiry, and together they form an interactive continuum.

9

Interviewing Method

Interviewing is widely used in qualitative research. Compared with observation, it is more economical in time, but may achieve less in understanding the culture. The economy in time, however, makes ethnographic interviewing almost the most widely used technique in policy-related research. Interviewing is trying to understand what people think through their speech. There are different types of interviews, often classified by the degrees of control over the interview.

Along this line, we shall briefly introduce three types of interviewing: informal interviewing, unstructured interviewing, semi-structured interviewing, and formally structured interviewing. We shall also briefly introduce key-informant interviewing and focus groups which are specific types of ethnographic interviewing.

Qualitative research of course has no monopoly over interviewing. Interviewing is also frequently used in research of other traditions. The difference between ethnographic interviewing and interviewing in other traditions lies mainly in two areas: the interviewer-interviewee relationship and the aims of interviews. Ethnographic interviewees, or informants, are teachers rather than subjects to the researcher, they are leaders rather than followers in the interview. The major aim of the interview should not be seeking responses to specific questions, but initiating the informant to unfold data. Readers may find more detailed discussions about ethnographic interviewing in Spradley (1979) who provides perhaps the most insightful account of the subject. In-depth discussions about ethnographic interviewing can also be found in Bernard (1988), Patton (1990), Fetterman (1989) and Powney and Watts (1987).

INTERVIEWS

Interviews are among the most challenging and rewarding forms of measurement. They require a personal sensitivity and adaptability as well as the ability to stay within the bounds of the designed protocol. Here, I describe the preparation you need to do for an interview study and the process of conducting the interview itself.

The Role of the Interviewer

The interviewer is really the "jack-of-all-trades" in survey research. The interviewer's role is complex and multifaceted.

It includes the following tasks:

- *Locate and enlist cooperation of respondents*: The interviewer has to find the respondent. In door-to-door surveys, this means being able to locate specific addresses. Often, the interviewer has to work at the least desirable times (like immediately after dinner or on weekends) because that's when respondents are most readily available.
- *Motivate respondents to do good job*: If the interviewer does not take the work seriously, why would the respondent? The interviewer has to be motivated and has to be able to communicate that motivation to the respondent. Often, this means that the interviewer has to be convinced of the importance of the research.
- *Clarify any confusion/concerns*: Interviewers have to be able to think on their feet. Respondents may raise objections or concerns that were not anticipated. The interviewer has to be able to respond candidly and informatively.
- *Observe quality of responses*: Whether the interview is personal or over the phone, the interviewer is in the best position to judge the quality of the information that is being received. Even a verbatim transcript will not adequately convey how seriously the respondent took the task, or any gestures or body language that were evident.
- *Conduct a good interview*: Last, and certainly not least, the interviewer has to conduct a good interview! Every interview has a life of its own. Some respondents are motivated and attentive, others are distracted or disinterested. The interviewer also has good or bad days. Assuring a consistently high-quality interview is a challenge that requires constant effort.

Training the Interviewers

One of the most important aspects of any interview study is the training of the interviewers themselves. In many ways the interviewers are your measures, and the quality of the results is totally in their hands. Even in small studies involving only a single researcher-interviewer, it is important to organize in detail and rehearse the interviewing process before beginning the formal study.

Here are some of the major topics that should be included in interviewer training:

- *Describe the entire study*: Interviewers need to know more than simply how to conduct the interview itself. They should learn about the background for the study, previous work that has been done, and why the study is important.
- *State who is sponsor of research*: Interviewers need to know who they are working for. They — and their respondents — have a right to know not just what agency or company is conducting the research, but also, who is paying for the research.
- *Teach enough about survey research*: While you seldom have the time to teach a full course on survey research methods, the interviewers need to know enough that they respect the survey method and are motivated. Sometimes it may not be apparent why a question or set of questions was asked in a particular way. The interviewers will need to understand the rationale for how the instrument was constructed.
- *Explain the sampling logic and process*: Naive interviewers may not understand why sampling is so important. They may wonder why you go through all the difficulties of selecting the sample so carefully. You will have to explain that sampling is the basis for the conclusions that will be reached and for the degree to which your study will be useful.
- *Explain interviewer bias*: Interviewers need to know the many ways that they can inadvertently bias the results. And, they need to understand why it is important that they not bias the study. This is especially a problem when you are investigating political or moral issues on which people have strongly held convictions. While the interviewer may think they are doing good for society by slanting results in favour of what they believe, they need to recognize that doing so could jeopardize the entire study in the eyes of others.
- *"Walk through" the interview*: When you first introduce the interview, it's a good idea to walk through the entire protocol so the interviewers can get an idea of the various parts or phases and how they interrelate.

Explain respondent selection procedures, including:

- *Reading maps*: It's astonishing how many adults don't know how to follow directions on a map. In personal interviews, the interviewer may need to locate respondents who are spread over a wide geographic area. And, they often have to navigate by night (respondents tend to be most available in evening hours) in neighbourhoods they're not familiar with. Teaching basic map reading skills and confirming that the interviewers can follow maps is essential.

- *Identifying households*: In many studies it is impossible in advance to say whether every sample household meets the sampling requirements for the study. In your study, you may want to interview only people who live in single family homes. It may be impossible to distinguish townhouses and apartment buildings in your sampling frame. The interviewer must know how to identify the appropriate target household.
- *Identify respondents*: Just as with households, many studies require respondents who meet specific criteria. For instance, your study may require that you speak with a male head-of-household between the ages of 30 and 40 who has children under 18 living in the same household. It may be impossible to obtain statistics in advance to target such respondents. The interviewer may have to ask a series of filtering questions before determining whether the respondent meets the sampling needs.
- *Rehearse interview*: You should probably have several rehearsal sessions with the interviewer team. You might even videotape rehearsal interviews to discuss how the trainees responded in difficult situations. The interviewers should be very familiar with the entire interview before ever facing a respondent.
- *Explain supervision*: In most interview studies, the interviewers will work under the direction of a supervisor. In some contexts, the supervisor may be a faculty advisor; in others, they may be the "boss." In order to assure the quality of the responses, the supervisor may have to observe a subsample of interviews, listen in on phone interviews, or conduct follow-up assessments of interviews with the respondents. This can be very threatening to the interviewers. You need to develop an atmosphere where everyone on the research team -- interviewers and supervisors -- feel like they're working together towards a common end.
- *Explain scheduling*: The interviewers have to understand the demands being made on their schedules and why these are important to the study. In some studies it will be imperative to conduct the entire set of interviews within a certain time period. In most studies, it's important to have the interviewers available when it's convenient for the respondents, not necessarily the interviewer.

The Interview

So all the preparation is complete, the training done, the interviewers ready to proceed, their "kits" in hand. It's finally time to do an actual interview. Each interview is unique, like a small work of art (and sometimes the art may not be very good). Each interview has its own ebb and flow -- its own pace. To the outsider, an interview looks like a fairly standard, simple, prosaic effort.

But to the interviewer, it can be filled with special nuances and interpretations that aren't often immediately apparent. Every interview includes some common components. There's the opening, where the interviewer gains entry and establishes the rapport and tone for what follows. There's the middle game, the heart of the process, that consists of the protocol of questions and the improvisations of the probe. And finally, there's the endgame, the wrap-up, where the interviewer and respondent establish a sense of closure. Whether it's a two-minute phone interview or a personal interview that spans hours, the interview is a bit of theater, a mini-drama that involves real lives in real time.

Opening Remarks

In many ways, the interviewer has the same initial problem that a salesperson has. You have to get the respondent's attention initially for a long enough period that you can sell them on the idea of participating in the study. Many of the remarks here assume an interview that is being conducted at a respondent's residence. But the analogies to other interview contexts should be straightforward.

- *Gaining entry*: The first thing the interviewer must do is gain entry. Several factors can enhance the prospects. Probably the most important factor is your initial appearance. The interviewer needs to dress professionally and in a manner that will be comfortable to the respondent. In some contexts a business suit and briefcase may be appropriate. In others, it may intimidate. The way the interviewer appears initially to the respondent has to communicate some simple messages -- that you're trustworthy, honest, and non-threatening. Cultivating a manner of professional confidence, the sense that the respondent has nothing to worry about because you know what you're doing -- is a difficult skill to teach and an indispensable skill for achieving initial entry.
- *Doorstep technique*: You're standing on the doorstep and someone has opened the door, even if only halfway. You need to smile. You need to be brief. State why you are there and suggest what you would like the respondent to do. Don't ask -- suggest what you want. Instead of saying "May I come in to do an interview?", you might try a more imperative approach like " I'd like to take a few minutes of your time to interview you for a very important study."
- *Introduction*: If you've gotten this far without having the door slammed in your face, chances are you will be able to get an interview. Without

waiting for the respondent to ask questions, you should move to introducing yourself. You should have this part of the process memorized so you can deliver the essential information in 20-30 seconds at most. State your name and the name of the organization you represent. Show your identification badge and the letter that introduces you. You want to have as legitimate an appearance as possible. If you have a three-ring binder or clipboard with the logo of your organization, you should have it out and visible. You should assume that the respondent will be interested in participating in your important study — assume that you will be doing an interview here.

- *Explaining the study*: At this point, you've been invited to come in (After all, you're standing there in the cold, holding an assortment of materials, clearly displaying your credentials, and offering the respondent the chance to participate in an interview — to many respondents, it's a rare and exciting event. They hardly ever get asked their views about anything, and yet they know that important decisions are made all the time based on input from others.). Or, the respondent has continued to listen long enough that you need to move onto explaining the study. *There are three rules to this critical explanation*:
  1. Keep it short;
  2. Keep it short; and
  3. Keep it short!

  The respondent doesn't have to or want to know all of the neat nuances of this study, how it came about, how you convinced your thesis committee to buy into it, and so on. You should have a one or two sentence description of the study memorized. No big words. No jargon. There will be more than enough time for that later (and you should bring some written materials you can leave at the end for that purpose). This is the "25 words or less" description. What you *should* spend some time on is assuring the respondent that you are interviewing them confidentially, and that their participation is voluntary.

Asking the Questions

You've gotten in. The respondent has asked you to sit down and make yourself comfortable. It may be that the respondent was in the middle of doing something when you arrived and you may need to allow them a few minutes to finish the phone call or send the kids off to do homework.

Now, you're ready to begin the interview itself:

- *Use questionnaire carefully, but informally*: The questionnaire is your friend. It was developed with a lot of care and thoughtfulness. While you have to be ready to adapt to the needs of the setting, your first instinct should always be to trust the instrument that was designed. But you also need to establish a rapport with the respondent. If you have your face in the instrument and you read the questions, you'll appear unprofessional and disinterested. Even though you may be nervous, you need to recognize that your respondent is most likely even more nervous. If you memorize the first few questions, you can refer to the instrument only occasionally, using eye contact and a confident manner to set the tone for the interview and help the respondent get comfortable.

- *Ask questions exactly as written*: Sometimes an interviewer will think that they could improve on the tone of a question by altering a few words to make it simpler or more "friendly." DON'T. You should ask the questions as they are on the instrument. If you had a problem with a question, the time to raise it was during the training and rehearsals, not during the actual interview. It is important that the interview be as standardized as possible across respondents (this is true except in certain types of exploratory or interpretivist research where the explicit goal is to avoid any standardizing). You may think the change you made was inconsequential when, in fact, it may change the entire meaning of the question or response.
- *Follow the order given*: Once you know an interview well, you may see a respondent bring up a topic that you know will come up later in the interview. You may be tempted to jump to that part of the interview while you're on the topic. DON'T. You are more likely to lose your place. You may omit questions that build a foundation for later questions.
- *Ask every question*: Sometimes you'll be tempted to omit a question because you thought you already heard what the respondent will say. Don't assume that. For example, let's say you were conducting an interview with college age women about the topic of date rape. In an earlier question, the respondent mentioned that she knew of a woman on her dormitory floor who had been raped on a date within the past year. A few questions later, you are supposed to ask "Do you know of anyone personally who was raped on a date?" You already know that the answer is yes, so you decide to skip the question. Instead, you might say something like "I know you may have already mentioned this, but do you know of anyone personally who was raped on a date?" At this point, the respondent may say something like "Well, in addition to the woman who lived down the hall in my dorm, I know of a friend from high school who experienced date rape."

- *Don't finish sentences*: I don't know about you, but I'm one of those people who just hates to be left hanging. I like to keep a conversation moving. Once I know where a sentence seems to be heading, I'm aching to get to the next sentence. I finish people's sentences all the time. If you're like me, you should practice the art of patience (and silence) before doing any interviewing. Silence is one of the most effective devices for encouraging a respondent to talk. If you finish their sentence for them, you imply that what they had to say is transparent or obvious, or that you don't want to give them the time to express themselves in their own language.

Obtaining Adequate Responses - The Probe

OK, you've asked a question. The respondent gives a brief, cursory answer. How do you elicit a more thoughtful, thorough response? You probe.

- *Silent probe*: The most effective way to encourage someone to elaborate is to do nothing at all - just pause and wait. This is referred to as the "silent" probe. It works (at least in certain cultures) because the respondent is uncomfortable with pauses or silence. It suggests to the respondent that you are waiting, listening for what they will say next.
- *Overt encouragement*: At times, you can encourage the respondent directly. Try to do so in a way that does not imply approval or disapproval of what they said (that could bias their subsequent results). Overt encouragement could be as simple as saying "Uh-huh" or "OK" after the respondent completes a thought.
- *Elaboration*: You can encourage more information by asking for elaboration. For instance, it is appropriate to ask questions like "Would you like to elaborate on that?" or "Is there anything else you would like to add?"
- *Ask for clarification*: Sometimes, you can elicit greater detail by asking the respondent to clarify something that was said earlier. You might say, "A minute ago you were talking about the experience you had in high school. Could you tell me more about that?"
- *Repetition*: This is the old psychotherapist trick. You say something without really saying anything new. For instance, the respondent just described a traumatic experience they had in childhood. You might say "What I'm hearing you say is that you found that experience very traumatic." Then, you should pause. The respondent is likely to say something like "Well, yes, and it affected the rest of my family as well. In fact, my younger sister..."

Recording the Response

Although we have the capability to record a respondent in audio and/ or video, most interview methodologists don't think it's a good idea. Respondents are often uncomfortable when they know their remarks will be recorded word-for-word. They may strain to only say things in a socially acceptable way.

Although you would get a more detailed and accurate record, it is likely to be distorted by the very process of obtaining it. This may be more of a problem in some situations than in others. It is increasingly common to be told that your conversation may be recorded during a phone interview. And most focus group methodologies use unobtrusive recording equipment to capture what's being said. But, in general, personal interviews are still best when recorded by the interviewer using pen and paper.

Here, I assume the paper-and-pencil approach:

- *Record responses immediately*: The interviewer should record responses as they are being stated. This conveys the idea that you are interested enough in what the respondent is saying to write it down. You don't have to write down every single word -- you're not taking stenography. But you may want to record certain key phrases or quotes verbatim. You need to develop a system for distinguishing what the respondent says verbatim from what you are characterizing (how about quotations, for instance!).
- *Include all probes*: You need to indicate every single probe that you use. Develop a shorthand for different standard probes. Use a clear form for writing them in (*e.g.*, place probes in the left margin).
- *Use abbreviations where possible*. Abbreviations will help you to capture more of the discussion. Develop a standardized system (*e.g.*, R=respondent; DK=don't know). If you create an abbreviation on the fly, have a way of indicating its origin. For instance, if you decide to abbreviate Spouse with an 'S', you might make a notation in the right margin saying "S=Spouse."

Concluding the Interview

When you've gone through the entire interview, you need to bring the interview to closure.

Some important things to remember:

- *Thank the respondent*: Don't forget to do this. Even if the respondent was troublesome or uninformative, it is important for you to be polite and thank them for their time.

- *Tell them when you expect to send results*: I hate it when people conduct interviews and then don't send results and summaries to the people who they get the information from. You owe it to your respondent to show them what you learned. Now, they may not want your entire 300-page dissertation. It's common practice to prepare a short, readable, jargon-free summary of interviews that you can send to the respondents.
- *Don't be brusque or hasty*: Allow for a few minutes of winding down conversation. The respondent may want to know a little bit about you or how much you like doing this kind of work. They may be interested in how the results will be used. Use these kinds of interests as a way to wrap up the conversation. As you're putting away your materials and packing up to go, engage the respondent. You don't want the respondent to feel as though you completed the interview and then rushed out on them — they may wonder what they said that was wrong. On the other hand, you have to be careful here. Some respondents may want to keep on talking long after the interview is over. You have to find a way to politely cut off the conversation and make your exit.
- *Immediately after leaving — write down any notes about how the interview went*: Sometimes you will have observations about the interview that you didn't want to write down while you were with the respondent. You may have noticed them get upset at a question, or you may have detected hostility in a response. Immediately after the interview you should go over your notes and make any other comments and observations — but be sure to distinguish these from the notes made during the interview (you might use a different colour pen, for instance).

TECHNIQUES SEMI-STRUCTURED INTERVIEWS

In an interview, what you already know is as important as what you want to know. What you want to know determines which questions you will ask. What you already know will determine how you ask them. Thanks to past jobs as a journalist and as an anthropological researcher, I've had training in both journalistic and ethnographic styles of interviewing. The two are at the opposite ends of the interview continuum. The journalistic style tries to verbally pin the respondent down by appearing to know everything already. The questions are direct and directed towards a particular outcome.

The ethnographic style of interviewing instead tries to enter into the world of the respondent by appearing to know very little. There are many types of interviews with many styles of questions, each appropriate in different circumstances. Unstructured interviews, often used by ethnographers, are really more conversations than interviews, with even the topic of conversation subject to change as the interview progresses. These "soaking and poking" experiences are most appropriate when the interviewer has limited knowledge about a topic or wants an insider perspective.

But the tendency for such interviews to wander off in unexpected directions-although they may provide for fresh ideas-almost guarantees that the interviews will not be a very consistent source of reliable data that can be compared across interviews.

Unstructured interviews are best used as a source of insight, not for hypothesis testing. Sometimes, however, we already have a lot of knowledge about a topic and want very specific answers to very specific questions.

When the researcher already knows a lot about the subject matter-the categories and all possible responses are familiar, and the only goal is to count how many people fall into each category of response-structured interviews with closed-ended questions are most appropriate.

Political scientists are most familiar with this type of interview because of mass public opinion surveys. Such closed-ended approaches can sometimes backfire, however, if we assume we are familiar with an area but end up asking the wrong questions in the wrong way or omitting an important response choice. We may find ourselves with reliable data that lacks any content validity.

There is a middle ground, however, and one that can provide detail, depth, and an insider's perspective, while at the same time allowing hypothesis testing and the quantitative analysis of interview responses. In this chapter, as suggested, focus on that interview style-semistructured interviews with open-ended questions. My observations and suggestions here come not only from my past experiences as a journalist and an ethnographic researcher, but also from my current research among lobbyists and policymakers in Washington, DC, as part of the Advocacy and Public Policymaking project.'

Gaining Rapport

Without rapport, even the best-phrased questions can fall flat and elicit brief, uninformative answers. Rapport means more than just putting people at ease. It means convincing people that you are listening, that you understand and are interested in what they are talking about, and that they should continue talking. There are several ways of doing this within the interview.

Putting Respondents at Ease

Some interviewing textbooks recommend that the interviewer "appear slightly dim and agreeable" or "play dumb" so that respondents do not feel threatened and are not womed that they will lose face in the interview. The danger here is that—especially when dealing with highly educated, highly placed respondents-they will feel that they are wasting their time with an idiot, or at least will dumb-down their answers and subject interviewers to a Politics 101 lecture. At the same time, the concern about respondents' feelings is valid. Even highly educated, highly placed respondents do not want to appear stupid in front of a university professor. I recommend a middle road.

The interviewer should seem professional and generally knowledgeable, but less knowledgeable than the respondent on the particular topic of the interview. So for me, I know a lot about lobbying and a lot about American politics, and I know what has been in the newspaper on a given policy issue, but I present myself as having little or no idea about what happened behind the scenes in the given policy issue I am interviewing about. I try to continue this approach even after I have conducted many interviews on the same policy issue. I don't want someone to leave something out because they assume I already knew it. A second, related point to remember is that many of your subjects will be more nervous than you are. After all, after you've done a couple of these, you are experienced. Your respondents, however, may never have beenthe subject of an academic study before. Reassure them by being open and avoiding threatening descriptions of your work. "Talk with you" is less threatening than "interview you," for example.

It is possible to be honest without being scary. There's no need to make your work sound like a medical procedure. Approach interview subjects with a positive attitude. Act as though it is natural that people would *ant* to talk to you. Appear friendly and curious. An important way to make an interview subject feel at ease is to explain your project again. This is the one-minute version of your project, which should describe the topic you are interested in and the types of questions you will ask, without tipping your hand as to your hypotheses.' At this point you can remind respondents that their answers are confidential.'

Are You Listening

During the interview itself, before moving on to the next question it often helps to briefly restate what the respondent has just said. This shows that you are interested and have understood what the respondent has just said. It also provides a chance for the respondent to correct you if you have misunderstood. Avoid *reinterpreting* what the respondent has just said, as this has the tendency to work against rapport and leave the respondents feeling as if the interviewer is trying to put words in their mouths. Use the respondents' own language, if possible, to summarize what has just been said.

Anthropologist James Spradley suggests that when an interviewer does not understand a particular point. that it is better to ask for *use* rather than *meaning*. That is, "When would you do that?" or "What would you use that for?" are usually better questions for building rapport than "What do you mean by that?" The latter tends to shift respondents out of their own verbal word and to begin speaking to you as an outsider.

Question Order

Question order is important for substantive reasons but order is also important as a means of gaining rapport. As any journalist would tell you, in an interview you should always move from the nonthreatening to the threatening. That is, ask the easy questions first. In the interviews I have conducted among lobbyists and policymakers, I find that it usually works better to ask things like age, background, title, and other personal things last. That way the interview doesn't come off as if it is about my respondent personally, but rather about the political issue or organization that we are talking about.

This type of question order works for me because my other questions are not personal, and are therefore even less threatening than the demographic information I collect at the end. On the other hand, if your interview questions focus on an individual's own political and philosophical beliefs, then obviously questions about education, background, and title would be less threatening and would provide a good place to start. When and how should you ask sensitive questions? It's usually best to wait until the middle or towards the end of the interview. Don't wait until the last minute-you may run out of time. Don't hem, haw, or make it seem as though any normal person would refuse to answer this question. Just ask. Then be quiet, and give the respondent time to answer. Most people will try to fill the silence, and you will get your answer.

A second thing to remember about sensitive questions-or any question, for that matter-is to use nonjudgmental, nonthreatening wording. For instance, asking a respondent, "What kinds of help do you give to members of Congress as they are going about their work or daily lives'?' is likely to gain you more information than if you were to ask, "Do you do favours for members of congress?" Likewise, I know that nonprofit organizations with 501charitable status are skittish about the word "lobbying," since the IRS restricts the amount of lobbying they can do. So I make a habit of referring to "advocacy efforts" or "policy work" instead.

When did you Stop Beating your Wife

What are known as "presuming" questions are common in journalism, but are usually not good social science. There are circumstances, however, when such questions are necessary to make respondents comfortable enough to answer honestly. When the question is one that the respondent is likely to try to avoid and involves a matter that may have a stigma attached to it, a presuming question may be the only way to go. When I was working as an ethnographic researcher in Kenya, collecting reproductive histories from women, I first began simply by asking women to tell me about all of their pregnancies. It was clear from the first few interviews that no one was mentioning miscarriages, stillbirths, or deaths of children-and I knew that could not be accurate in a rural area with nonexistent prenatal care and high child mortality.

So I tried probing: "Tell me about any children who died." I used this question only once, and it caused a respondent to jump up, mutter that she must go check on the goats, and run out the door. After some help from a language consultant, I did two things. I made my language less threatening, and I asked the question in a presuming way. "How many children are the lost ones?" I asked-"Aja inkera netala?" My respondents' faces would turn serious, they would sigh, then they would tell me the details I was seeking. To return to the political world, instead of asking a lobbyist, "Did you give soft money donations?" it might make the question easier to answer to say, "How much did your organization give in soft money donations?'The latter presumes that it is normal to give soft money donations and that everyone must do it, and also shifts the onus away from the individual and onto the organization. Presuming questions are presuming in the sense that they imply that the researcher already knows the answer—or at least part of it. So one danger is that the respondent will bluff to save face and make something up. That is why I suggest such questions should be used very sparingly and only when they are needed to take the edge off of questions that may otherwise have a stigma attached. A respondent would be relieved, not shamed to be able to say "None."

Types of Questions

We all know that there are certain types of questions to avoid-loaded questions, double-barreled questions, leading questions, and presuming questions. But what types of questions *should* you ask in an open-ended, semistructured interview?

Grand Tour Questions

The single best question *I* know of for a semistructured interview is what Spradley calls *a grand tour question.* Like the name suggests, these questions ask respondents to give a verbal tour of something they know well. The major benefit of the question is that it gets respondents talking, but in a fairly focused way. Many good interviewers use this type of question instinctively. Jeff Berry, for example, used one in the research for his book *The Interest Group Society* when he asked lobbyists to describe an average day. There are many different types of grand tour questions.

The most common is probably the typical grand tour question:

- "Could you describe a *typical* day in your office?"
- "Could you describe a *typical* day on the Hill?"
- "Could you describe a *typical* day in a member of Parliament's office?"

Such questions have the benefit of giving you a sense of what an average day is like, but the drawback that you are not certain what is being averaged-that is, how much variation there is and how accurate the respondent's sense of the usual really is. Respondents may have a tendency to focus on the interesting, or on what they think *should* happen day to day. If you are doing enough interviews to get a sense of the average by comparing across interviews, then you may want to turn to a *specific grand tour* question. Specific grand tour questions ask for a tour based on some parameter decided by the interviewer-a day, a topic, an event: "Could you walk me through what you did *yesterday* in your office?'or "Walk me through what your organization did *in response to issue X.*"

We used a specific grand tour question to begin our interviews for the Advocacy and Public Policymaking project, asking respondents to describe their organizations' activities on the *most recent* policy issue in which they were involved. Not all interviews need to be conducted sitting down, and not all grand tours need to be virtual. A *guided grand tour* is an actual tour: "The next time you are lobbying on the Hill, could you bring me along and show me what you do?Related to this are *task-related grand tours.* Such questions ask the respondent to perform some usual task while verbally walking the interviewer through the task. For instance, I could ask a lobbyist to lay out talking points for a meeting with a legislator, or to compile a list of which members of Congress to talk to, explaining the decisions being made at each step of the process.

Prompts

Prompts are as important as the questions themselves in semistructured interviews. Prompts do two things: they keep people talking and they rescue you when responses turn to mush.

Let's take the introductory question from the Advocacy and Public Policymaking project:

- "Could you take the most recent issue you've been spending time on and describe what you're trying to accomplish on this issue and what type of action are you taking to make that happen?"

One of my respondents answered, "Well, we've been talking to some people on the Hill and trying to get our message out." He had just described the activities of every lobbyist in Washington. If I had stopped here, the interview would have been useless. Luckily, my interview protocol included numerous prompts, based on what we wanted to be able to code from this question, including who the targets of lobbying were, and what lobbying tactics were used. So at this point, possible prompts would include: "Who have you been talking to on the Hill?" and "What are you doing to try to get your message out?" McCracken identifies several different types of prompts. Prompts like the ones I just mentioned are planned prompts-prompts that are formally included in the interview protocol. At the end of each formal question we ask as part of the Advocacy and Public Policymaking project, there is an italicized list of specifics that the interviewer is supposed to probe for if the respondent doesn't bring them up.

For example:

- Probe about coalition partners probe about who they are speaking with about this issue

One difference between a prompt and a question is that the prompts are not scripted as are the initial questions. The reason is that every interview is different and the list of possible probe situations could potentially go on for dozens of pages. That makes it important for the interviewer to have a plan for how the interviews will eventually be coded, so that the interviewer can make sure that the responses have covered the necessary points. Probably the most instinctive type of prompt is an informal prompt. This is an unscripted prompt that may be nothing more than the reassuring noises and interjections that people make during any conversation to show that they are listening and interested: "Uh-huh." "Yes." "How interesting." But the welltrained interviewer has a variety of informal prompts to use. Floating prompts, for example, are used to clarify. These may be nothing more than raising an eyebrow and cocking one's head, or they may be specific questions: "How?" "Whv?" and "And then...?"

One wav to ask for clarification and at the same time build rapport is to repeat the key term of the respondent's last remark as a question:

- *Respondent:* "And the bill was completely whitewashed in committee."
- *Interviewer:* "Whitewashed?"

McCracken warns against leading respondents by putting words in their mouths You risk losing rapport or having the respondent go along with your definition, rather than clarifying further. The goal here is to listen for key terms and to prompt the respondent to say more about them.

Enough is Enough

One of the most important rules about asking questions has to do with shutting up. Give your respondent room to talk. If respondents get off topic, let them finish, then bring them gently back to the issue you are interested in. But don't try to control too much or you may miss important, unexpected points.

INFORMAL INTERVIEWING

Informal interviewing entails no control. It is usually conversations that the researcher recall after staying in the field. It is different from "observation" in that it is interactive. That is, the informant speaks to the researcher. By its own nature, informal interviewing is the most "ethnographic" in the sense that it is not responding to any formal question. It is part of the self-unfolding process.

Conversation during Travels

Almost without exception, doing research in rural China involves long travels, usually one land, sometimes by sea. Typically, the whole group of six or seven travels in a van. The van may travel over five or six hours, or more. The journeys are usually full of conversations.

When it is too long, people tend to take naps, but still, between naps people talk. The conversations often provide valuable information not available during more formal sessions. For example, I had always thought that the crucial factor that led to the low income of community-supported teachers (Case 6) was finance.

I was only through conversation that I realized it was more a matter of rural-urban disparity in citizenship. As another exame, once while was had a break on the road and stood by the van. I said jokingly, "what do you think if the province were a country on its own". This evoked emotions conversations which gave me a vivid description of the local planners' perception of decentralization. A third example: I slept overnight on a boat when I was travelling in a province to a small town. The man in the same cabin happened to be a senior official in the local finance department. He told me all kinds of reforms and shortcomings in the local financial system which help me grasp an alternative perspective of educational finance.

Unstructured Interviewing

Unstructured interviewing applies minimum control over the informant and the responses. Unstructured interviewing is formal interviewing. There is no disguise that the occasion is just a friendly chat. Hence, there is always a question to answer, or a topic to discuss. However, the informants have the liberty to choose their own scope, depth, emphases, length and pace of the response. At times, the informant may even choose to deviate from the original question or topic.

Training Needs of Supervisors

Sung (1992) attempted to study the management training needs of supervisors of child-care centres of the disabled, in the view of designing appropriate training programmes for the supervisors. She started with ethnographic interviewing with some of the centre supervisors. The interviews started with the supervisors perceptions about "problems" in their work. The supervisors were quite articulate of their problems. However, none of the first few informants mentioned anything about management as their felt problem, nor did they relate their problems with training. In other words, the interviews did not answer Sung's original question about training needs. However, the interviews proved enlightening, because they provided the actual perspectives of the supervisors. For example, they are more aware of their roles as programmes developers, and are less conscious of their managerial roles. Some of them in fact wished they could get rid of their managerial tasks. This allowed Sung to have a completely fresh look at the meaning of training of centre supervisors.

Most of what is known as "ethnographic interviews" are unstructured. This is especially the case at the early stages of the research, when the researcher knows little about the subject to be studied and when there is comparatively plenty of time. As the research goes on, the researcher may have gradually developed his/her thinking and may choose to ask more specific questions in a narrower scope. The interviews then tend to be more structured.

Participation in Educational Policy-making

In 1983, I did a research which attempted to identify the patterns of participation by educational bodies in educational policy-making in Hong Kong. The main body of the research was a questionnaire which asked for each educational body's nature and its mode of participation in government policy-making. the 68 educational bodies which provide valid data were then classified according to their modes of participation by cluster analysis and factor analysis. It was basically a quantitative analysis. The design of the questionnaire was based on 14 semi-structured interviews with people who had extensive experience with educational bodies. The design of these interviews were in turn based on two background interviews with the Senior Assistant Director of Education (who was the *de facto* chief in planning) and the most prominent union leader. Both interviews were unstructured and started with the grand tour question "What do you see as the picture of participation in educational planning in Hong Kong?"(Cheng, 1983:68). The informants were then allowed to elaborate according to what they knew about the topic. There was no set time limit for the interviews.

Unstructured interviewing is often characterized by the very few questions asked by the researcher. It is therefore sometimes also called the open-ended interviewing. Indeed, a successful qualitative unstructured interview often starts with a "grand tour" question and no further question is necessary until the informant has said all that is to be said about that topic. A grand tour question is designed to elicit a broad picture of the native's world.

Grand tour questions should lead the researcher to understand the general framework in which the informants think, the terms they used, the context in which such terms are used, and so forth. The grand tour question also gives the researcher a basis to frame further questions. In practice, the researcher often has to do some "probing" in order to continue the conversation. Skill is required in such probing so that it widens rather than narrows the scope of conversion.

Unstructured interviewing often encounters information that the researcher is unprepared for. Such information are valuable input which may change the researchers' perceptions. The following is an example.

Farmers send their children to schools

In an IIEP project in Zhejiang province of China, one of the research objective was to understand why parents sent their children to schools.

We started with a crude questionnaire. One item in the questionnaire was the parent's expected future career of their children. The questionnaire, census categories were used. In addition, some experienced teachers with good knowledge of the past graduates helped check the list against all the possible occupations past primary graduates had taken up. The section ended up with the following occupations:

(01) further studies

(02) administrator in the village

(03) professional technicians

(04) skilled workers
(05) professional farmer
(06) businessman
(07) local industrialist
(08) military personnel
(09) teacher
(10) others.

However, ethnographic interviews with parents in a pilot village proved that the questionnaire was far from painting the real picture. Rural parents were often puzzled by the question "Why are you sending your children to schools?" The most articulate of them expressed their puzzle in a retorting question: "Why not?" or "Why should this be a question?". We realized that these responses were valuable data. We then asked: "When your child has completed primary school, what would you expect him/her to do?" The question did not normally get at immediate answers. Many parents, particularly the illiterates, reacted with doubts and fussed about the question. We took these responses as valuable data. The following are some of such responses:

"How should I know?"

"What do you mean?"

"What do you think I should say?" (Turning to the local teacher.)

"I don't know how to answer."

"I have never thought of it!"

We identified that there were parents who actually did not think much about their children's future. This may or may not reflect a total neglect of their children, as the following answers to further probing indicate:

"It goes without saying that I wish they could go to universities."

"Of course I want them to get education as high up as possible, but I don't know what they are called."

"I don't mind, as far as they can leave the village eventually."

"I don't mind what they do, as long as they are not doing anything bad."

In the end, we decided that the interviewers should ask the question in an open-ended mode and allow answers such as

"I have never thought of that"

"university study"

"further study, as high as possible"

"leave the village eventually"

"as far as not doing anything bad, irrespective of the occupation"

The actual results of the questionnaire over a sample of over 200 parents revealed that most of these items gained highest scores among rural parent.

SEMI-STRUCTURED INTERVIEWING

Semi-structured interviewing is interviewing with an interview guide. Semi-structured interviewing follows all the principles of unstructured interviewing, except that the informants are not expected to move too far beyond the scope defined by the interview guide. By the nature of qualitative research, however, the researcher tends not to stop the informant when the interview goes beyond the designed scope. Nevertheless, the guide at least allows the researcher to obtain data within the designed scope. This is useful in situations where the researcher can only meet the informant once and hence the interview is not likely to repeat.

Overseas Informants

When I carried out the study of educational policy-making in Hong Kong , many of the informants were interviewed overseas. One of the informants was an expert planner in Bonn, Germany who was once a consultant of the Hong Kong government. I interviewed him when I was able to travel to Bonn by train after a meeting in Paris *en route* to London. There were only two hours. I sent in advance a page of four broad questions about one policy-episode which he participated. Each question was preceded by a small paragraph to remind him of the context.

The questions were open-ended. The questions helped very much. The informant was able to recollect his experience in Hong Kong before I went. The informant in fact *led* the interview. Another informant was a chief educational administrator in Canberra, Australia.

I was able to spend two days in the city. Again, a similar page of broad questions were sent in advance. I asked for two visits on the two consecutive days. The first day interview was a grand tour of his perception of the particular episode in focus. I made a study of the interview results and framed questions for the second day interview. A third informant was a retired official from the Treasury of Britain. He was one of the first members of the Hong Kong University Grants Committee which was one of my foci of attention. In the letter I sent in advance, some broad questions about asking him to enlighten me with his perceptions of the background, status and role of the Committee. He was so pleased about the interview that it went from 1 p.m. until 6 p.m. in the evening. He invited me to a second interview, which took another four hours. He was not always focusing on the Committee. He started with the beginning of the British counterpart (UGC) in 1919, traced its development, and related its features to the Hong Kong Committee. He made comments not only on the committees and the policies, but also on persons. I was not at all worried by the "out-of-focus" conversation, and anyway there was no way to stop him. The interview gave me a splendid picture of not only the policies and organizations under research, but also dynamics and culture within the organizations.

The broad questions in the first two cases played the role of an interview guide. They helped to economize the limited time available in such interviews. The broad questions in the third case played a different role. They helped initiate the informant in unfolding what was in his mind. Hence, the intended semi-structure in the third case gave way to an unstructured interview, quite pleasantly for a qualitative researcher.

Structured Interviewing

In structured interviewing, all informants are asked to respond to the same set of questions or stimuli, or, to as nearly identical a set of stimuli as possible. Structured interviewing often involves a rather elaborate interview schedule. Hence, structured interviewing is not very different from using a questionnaire. In fact, Bernard (*Ibid.*) regards even self-administered questionnaire as a special type of structured interviewing. However, there remains a variety of approaches in structured interviewing. It still allows the use of ethnographic techniques to various degrees.

Structured Questionnaire with Ethnographic Interviewing

During the IIEP research on basic education, parents were interviewed with a questionnaire. The researcher took the questionnaire and interviewed parents many of whom were illiterate. The large number of parents, the considerable number of researchers, the numerous items and various possible responses made the questionnaire necessarily structured, with as many options possible for each item.

The questionnaire was printed and the researcher used one questionnaire for each parent. During the designing process, we faced two options: the researcher might either read out the optional responses for the informant to select, or the researcher might stop at asking the question and allow the respondent to give the answer. Eventually, we decided to adopt the latter, that is, allowing the parents to provide the answer. The reason was quite obvious. We discovered during the design of the questionnaire that parents' answers often went beyond our expectations. Although we had included as many as possible of these responses in the final draft of the questionnaire, we could never be sure that other parents would not have other answers in mind. If we asked the parents to respond to optional answers, we might still limit them to what we had expected and miss those unexpected.

The above is just a brief discussion of using interviewing in qualitative research. Readers may find more detailed discussions in McCracken (1988),

Bernard (1988) and Powney (1987). Spradley (1979) in particular provides a very useful systematic analysis of interviewing in the qualitative conventions.

Focus Group Interviewing

In the broad sense, focus group interviewing is in-depth group interviewing. Although the method itself may be used in all conventions of research, it tends to generate qualitative data in the *emic* sense (i.e. in the participants own words and own categories). Hence, focus group interviewing is often group interview in the ethnographic sense.

Typically, a group of 8 to 12 informants are invited to a session of around one and a half to two and a half hours (*Ibid.*:10). The researcher initiates discussion by whatever way appropriate, such that the participants are motivated but not guided to discuss on the relevant topic. The researcher will then listen and observe the discussions that follow, applying as little intervention as possible, as long as the discussion remains on the topic. The participants are expected to unfold their knowledge and express their opinions about the subject matter. Focus group interviewing is also known as focus group discussion, perhaps because in ideal situations, interaction between participants prevails over "interviewing" in its conventional sense.

The strengths of focus group interviewing are at least twofold. First, a large amount of information can be released in a comparatively short period of time. Second, it also allows the researcher to observe interactions among participants. Such interaction are not observable in individual interviews.

The major weakness of focus group interviewing is that under group situations, group dynamics and group pressure come into play. Participants may not be ready to express their independent views, or they may be actually influenced by the prevalent views during the discussion. The information obtained may well be distorted information. Such a weakness is perhaps less significant in a Western community than in a confucian society such as those in East Asia. In the former, individuals stand as independent entities and participants are more ready to accept diverse views. In the latter, people often believe in collectivism and correctness of views and what is expressed in groups may differ significantly from what each individual think.

A focus group interview is more successful when the topic is the participants' concern rather than only the researcher's concern. In this ideal case, the discussion is usually very focused with little intervention from the researcher, who acts as the moderator of discussion. In other cases, where the topic of discussion is not so much the concern of the participants, the research may have to apply more guidance during the discussion and the ethnographic sense is preserved only with caution.

Community Perception of Basic Education

A focus group discussion was held with a Neighbourhood Committee during a research project to under the environments of basic education in China. Eight members were present. The members were elected from the families who lived in the neighbourhood which was made up of a street. My concern was to understand what are perceived as the basic education needs, so that the questionnaire we were to design would be consistent with the community's perceptions. In order to focus the discussion, I started with an open question: "What are the characteristics of a man who cannot survive the society?" The responses converged on two points: good adaptability and good human relations. Adaptability includes adaptation to the nature and changing environments, endurance and persistence and self-study abilities. Good human relations include relations with family members, relations with peers and work-mates, and the management of self in an organization. Those who cannot survive were those who were unable to cope with the changing environment, who evaded difficulties, who were unable to manage themselves and those who tended to be self-isolating. No one mentioned literacy and numeracy which were our anticipated answers.

We followed with a number of probing questions each relating to literacy and numeracy. "Don't you think a young person should learn to change an electric bulb?" "Don't you think a young person should be able to read the road sign?" The participants did not seem to appreciate that these should become part of schooling. "They can always learn from their neighbour or even ask the neighbour to help." "They can always ask passers-by."

The focus group discussion above proved extremely helpful. It was an international comparison and we had always been focusing on literacy and numeracy as the basics of basic education. The attainment tests turned out that most students had achieved the basic requirements between primary 4 and primary 6. Parents therefore took literacy and numeracy for granted and turned to social requirements. The picture of what basic education was meant to parents was totally changed. This subsequently helped interpret a large number of research findings which might otherwise became inexplicable.

School Needs for Teacher Development

The Advisory Committee on Teacher Education and Qualification found it necessary to know how teacher education and teacher qualifications are matching the needs of schools. It was decided that a survey would be conducted. The survey could be conducted in the form of a questionnaire asking schools about the appropriateness of various activities in teacher education and development.

This, however, may narrow the respondents' views to what is existing, or what is perceived as ideal in the teacher educators' perspectives. The working group eventually decided to conduct two focus group discussions, one for primary schools and one for secondary schools. In each group, knowledgeable teachers and principals are invited to articulate their views about what they perceive of the most essential needs in schools. They are asked to limit their views on teachers.

The focus groups proved extremely useful in providing insights about the basic issues. Such issues, such as the non-teaching workload of teachers, and the structural responsibility of teacher development in schools, are beyond teachers' education and qualification, but are essential to solve teacher problems.

The richness in data obtain from focus group discussions may also mean difficulties in analyzing such data. This is different from individual interviews where the informant follows a certain line of thinking. In focus group discussions, the relations between views and people are complex. Such a complexity, one may argue, is indeed a more accurate reflection of the reality.

10

Quantitative Techniques in Management

Quantitative Techniques are those statistical and operation research techniques which help in the decision making process especially concerning business and industry. These techniques involve the introduction of the elements of quantities i.e. they involve the use of numbers, symbols and other mathematical expressions.

Two major divisions in which we can divide these techniques are-

1. Statistical Techniques (or Statistical methods and measures)
2. Programming Techniques (or Operations Research) Statistical technique.

A type of mathematical analysis involving the use of quantified representations, models and summaries for a given set of empirical data or real world observations. Statistical analysis involves the process of collecting and analyzing data and then summarizing the data into a numerical form.

Statistics is a general term used to summarize a process that an analyst, mathematician or statistician can use to characterize a data set. If the data set is based on a sample of a larger population, then the analyst can extend inferences onto the population based on the statistical results from the sample.

Some statistical measures include regression analysis, mean, kurtosis, skewness, analysis of variance and variance.

PROGRAMMING TECHNIQUE OR OPERATION RESEARCH

Operations Research, also known as **operational research**, is an interdisciplinary branch of applied mathematics and formal science that uses advanced analytical methods such as mathematical modelling, statistical analysis, and mathematical optimization to arrive at optimal or near-optimal solutions to complex decision-making problems. It is often concerned with determining the maximum (of profit, performance, or yield) or minimum (of loss, risk, or cost) of some real-world objective.

Operations research encompasses a wide range of problem-solving techniques and methods applied in the pursuit of improved decision-making and efficiency. Some of the tools used by operations researchers are statistics, optimization, probability theory, queuing theory, game theory, graph theory,decision analysis, mathematical modeling and simulation. Because of the computational nature of these fields, Operations Research also has strong ties to computer science. Operations researchers faced with a new problem must determine which of these techniques are most appropriate given the nature of the system, the goals for improvement, and constraints on time and computing power.

Work in operations research and management science may be characterized as one of three categories.

- Fundamental or foundational work takes place in three mathematical disciplines: Probability, Optimization, and Dynamical systems theory.
- Modeling work is concerned with the construction of models, analyzing them mathematically, implementing them on computers, solving them using software tools, and assessing their effectiveness with data. This level is mainly instrumental, and driven mainly by statistics and econometrics.
- Application work in operations research, like other engineering and economics' disciplines, attempts to use models to make a practical impact on real-world problems.

The major subdisciplines in modern operations research, as identified by the journal *Operations Research*, are:

- Computing and Information Technologies
- Decision Analysis
- Environment, Energy & Natural Resources
- Financial Engineering
- Manufacturing, Service Sciences, and Supply Chain Management
- Policy modeling and public sector work
- Revenue Management
- Simulation Stochastic Models
- Transportation

SCALE OF DATA MEASUREMENT

Before any statistical technique is employed, a researcher must determine the type of data that is to be collected. In a general sense, there are two types of data: qualitative data and quantitative data.

Qualitative data categorizes an element by a non-numeric attribute. For instance, if we are interested in which political party a resident belongs to, we are categorizing the resident using qualitative data: Democratic, Republican, Independent, etc. Qualitative data is often the data we are interested in gathering in the social sciences and particularly in business. For instance, much of what we want to know in business is related to attitudes or behavior of consumers. The data is not numeric and therefore more difficult to analyze. We often calculate the proportion of elements with a particular characteristic (i.e., the proportion of residents who own their own home) but many techniques cannot be used on this type of data.

There are **two types** of qualitative data: nominal data and ordinal data. **Nominal data** is, in terms of structure, the lowest form of data. Nominal data is qualitative data that has no natural order. Examples of nominal data include: gender; political affiliation; type of car owned; product model; etc. Data comprised of "numbers" can also be qualitative data. Zip codes, area codes, telephone numbers are examples of data that are qualitative. In math terms, these data are not "real" numbers because they do not represent **numeric measures**. One way to determine whether "numbers" are numeric measures is to consider whether one might be interested in an average of these "numbers". If a number can be replaced with letters, words or symbols without losing any information then this indicates that a "number" is not a numeric measure. **Ordinal data** is qualitative data that has a natural order. Examples of ordinal data include: military rank; size of clothing using S, M, L, XL; place in which a race was finished; condition of a used appliance using POOR, AVERAGE, GOOD, EXCELLENT; etc. While ordinal data has an order, the intervals between the rankings are not equal intervals. Thus, while ordinal data has more structure than nominal data, math functions on the data, such as differences, are not valid.

Quantitative data categorizes an element by a numeric measure. quantitative data are true numbers and, as a result, more quantitative techniques are available for use with this data. Quantitative data can be divided into **two types** of data: interval data and ratio data. **Interval** data is Quantitative data that has no natural starting point or zero level. Examples of interval data include Fahrenheit temperature and scores on IQ tests. Each (of these type data) is a numeric measure but neither has a natural starting point or zero level. Zero degrees Fahrenheit is not the absence of temperature just as there is no zero level for a test of intelligence. Interval data can be used for any technique that requires quantitative data, however, we must realize that ratios have no meaning with this type of data since there is no natural zero level. For example, 50 degrees Fahrenheit is not twice as warm as 25 degrees Fahrenheit. **Ratio** data is quantitative data that has a natural starting point or zero level. Most quantitative data falls into this scale of data measurement. Examples of ratio scaled data include height, weight, rate of return, net income, etc. Since there is a natural zero level, ratios have meaning.

MARKETING EFFECTIVENESS

Marketing effectiveness is the quality of how marketers go to market with the goal of optimizing their spending to achieve good results for both the short-term and long-term. It is also related to Marketing ROI and Return on Marketing Investment (ROMI).

Marketing effectiveness has four dimensions:

- Corporate – Each company operates within different bounds. These are determined by their size, their budget and their ability to make organizational change. Within these bounds marketers operate along the five factors described below.
- Competitive – Each company in a category operates within a similar framework as described below. In an ideal world, marketers would have perfect information on how they act as well as how their competitors act. In reality, in many categories have reasonably good information through sources, such as, IRI or Nielsen. In many industries, competitive marketing information is hard to come by.
- Customers/Consumers – Understanding and taking advantage of how customers make purchasing decisions can help marketers improve their marketing effectiveness. Groups of consumers act in similar ways leading to the need to segment them. Based on these segments, they make choices based on how they value the attributes of a product and the brand, in return for price paid for the product. Consumers build brand value through information. Information is received through many sources, such as, advertising, word-of-mouth and in the (distribution) channel often characterized with the purchase funnel, a McKinsey & Company concept. Lastly, consumers consume and make purchase decisions in certain ways.
- Exogenous Factors – There are many factors outside of our immediate control that can impact the effectiveness of our marketing activities. These can include the weather, interest rates, government regulations and many others. Understanding the impact these factors can have on our consumers can help us to design programs that can take advantage of these factors or mitigate the risk of these factors if they take place in the middle of our marketing campaigns.

There are five factors driving the level of marketing effectiveness that marketers can achieve:

1. Marketing Strategy – Improving marketing effectiveness can be

achieved by employing a superior marketing strategy. By positioning the product or brand correctly, the product/brand will be more successful in the market than competitors' products/brands. Even with the best strategy, marketers must execute their programs properly to achieve extraordinary results.

2. Marketing Creative – Even without a change in strategy, better creative can improve results. Without a change in strategy, AFLAC was able to achieve stunning results with its introduction of the Duck (AFLAC) campaign. With the introduction of this new creative concept, the company growth rate soared from 12% prior to the campaign to 28% following it.
3. Marketing Execution – By improving how marketers go to market, they can achieve significantly greater results without changing their strategy or their creative execution. At the marketing mix level, marketers can improve their execution by making small changes in any or all of the 4-Ps (Product, Price, Place and Promotion) (Marketing) without making changes to the strategic position or the creative execution marketers can improve their effectiveness and deliver increased revenue. At the program level marketers can improve their effectiveness by managing and executing each of their marketing campaigns better. It's commonly known that consistency of a Marketing Creative strategy across various media (e.g. TV, Radio, Print and Online), not just within each individual media message, can amplify and enhance impact of the overall marketing campaign effort. Additional examples would be improving direct mail through a better call-to-action or editing web site content to improve its organic search results, marketers can improve their marketing effectiveness for each type of program. A growing area of interest within (Marketing Strategy) and Execution are the more recent interaction dynamics of traditional marketing (e.g. TV or Events) with online consumer activity (e.g. Social Media). Not only direct product experience, but also any stimulus provided by traditional marketing, can become a catalyst for a consumer brand "groundswell" online as outlined in the book Groundswell.
4. Marketing Infrastructure (also known as Marketing Management) – Improving the business of marketing can lead to significant gains for the company. Management of agencies, budgeting, motivation and coordination of marketing activities can lead to improved competitiveness and improved results. The overall accountability for brand leadership and business results is often reflected in an organization under a title within a (Brand management) department.

5. Exogenous Factors-Generally out of the control of marketers, external or exogenous factors also influence how marketers can improve their results. Taking advantage of seasonality, interests or the regulatory environment can help marketers improve their marketing effectiveness.

RETURN ON MARKETING INVESTMENT

Return on Marketing Investment (ROMI) and Marketing ROI are defined as the optimization of marketing spend for the short and long term in support of the brand strategy by building a market model using valid, objective marketing metrics. Improving ROMI leads to improved marketing effectiveness, increased revenue, profit and market share for the same amount of marketing spend.

Metrics

There are two forms of the Return on Marketing Investment (ROMI) metric.

- short term ROMI
- long term ROMI.

Short Term

The first, short term ROMI, is also used as a simple index measuring the dollars of revenue (or market share, contribution margin or other desired outputs) for every dollar of marketing spend.

For example, if a company spends $100,000 on a direct mail piece and it delivers $500,000 in incremental revenue then the ROMI factor is 5.0. If the incremental contribution margin for that $500,000 in revenue is 60%, then the margin ROMI (the amount of incremental margin for each dollar of marketing spent is 3.0 (= 5.0 x 60%).

The value of the first ROMI is in its simplicity. In most cases a simple determination of revenue per dollar spent for each marketing activity can be sufficient enough to help make important decisions to improve the entire marketing mix.

Long Term

In a similar way the second ROMI concept, long term ROMI, can be used to determine other less tangible aspects of marketing effectiveness.

For example, ROMI could be used to determine the incremental value of marketing as it pertains to increased brand awareness, consideration or purchase intent. In this way both the longer term value of marketing activities (incremental brand awareness, etc.) and the shorter term revenue and profit can be determined.

This is a sophisticated metric that balances marketing and business analytics and is used increasingly by many of the world's leading organizations (Hewlett-Packard and Procter & Gamble to name two) to measure the economic (that is, cash-flow derived) benefits created by marketing investments. For many other organizations, this method offers a way to prioritize investments and allocate marketing and other resources on a formalized basis.

Criticism and Defence of Marketing Effectiveness

Direct measures of the short term variant of ROMI are often criticized as only including the direct impact of marketing activities without including the long-term brand building value of any communication inserted into the market.

Short term ROMI is best employed as a tool to determine marketing effectiveness to help steer investments from less productive activities to those that are more productive. It is a simple tool to gauge the success of measurable marketing activities against various marketing objectives (e.g., incremental revenue, brand awareness or brand equity). With this knowledge, marketing investments can be redirected away from under-performing activities to better performing marketing media.

Long term ROMI is often criticized as a 'silo-in-the-making"-it is intensively data driven and creates a challenge for firms that are not used to working business analytics into the marketing analytics that typically determine resource allocation decisions. Long term ROMI, however, is a sophisticated measure used by a number of forward thinking firms interested in getting to the bottom of value for money challenges often posed by competing brand managers.

COMMONLY USED MARKETING RESEARCH TERMS

Market research techniques resemble those used in political polling and social science research. Meta-analysis (also called the Schmidt-Hunter technique) refers to a statistical method of combining data from multiple studies or from several types of studies. Conceptualization means the process of converting vague mental images into definable concepts. Operationalization is the process of converting concepts into specific observable behaviours that a researcher can measure. Precision refers to the exactness of any given measure. Reliability refers to the likelihood that a given operationalized construct will yield the same results if re-measured. Validity refers to the extent to which a measure provides data that captures the meaning of the operationalized construct as defined in the study. It asks, "Are we measuring what we intended to measure?"

- Applied research sets out to prove a specific hypothesis of value to the clients paying for the research. For example, a cigarette company might commission research that attempts to show that cigarettes are good for one's health. Many researchers have ethical misgivings about doing applied research.
- Sugging (from "SUG", for *selling under the guise* of market research) forms a sales technique in which sales people pretend to conduct marketing research, but with the real purpose of obtaining buyer motivation and buyer decision-making information to be used in a subsequent sales call.
- Frugging comprises the practice of soliciting funds under the pretence of being a research organization.

SELECTING A RESEARCH SUPPLIER

A firm that cannot conduct an entire marketing research project in-house must select an external supplier for one or more phases of the project. The firm should compile a list of prospective suppliers from such sources as trade publications, professional directories, and word of mouth.

When deciding on criteria for selecting an outside supplier, a firm should ask itself why it is seeking outside marketing research support. For example, a small firm that needs one project investigated may find it economically efficient to employ an outside source. Or a firm may not have the technical expertise undertake certain phases of a project or political conflict-of-interest issues may determine that a project be conducted by an outside supplier.

When developing criteria for selecting an outside supplier, a firm should keep some basics in mind. What is the reputation of the supplier? Do they complete projects on schedule? Are they known for maintaining ethical standards? Are they flexible? Are their research projects of high quality? What kind and how much experience does the supplier have? Has the firm had experience with projects similar to this one? Do the supplier's personnel have both technical and nontechnical expertise? In other words, in addition to technical skills, are the personnel assigned to the task sensitive to the client's needs and do they share the client's research ideology? Can they communicate well with the client?

The cheapest bid is not always the best one. Competitive bids should be obtained and compared on the basis of quality as well as price. A good practice is to get a written bid or contract before beginning the project. Decisions about marketing research suppliers, just like other management decisions, should be based on sound information.

CAREERS IN MARKETING RESEARCH

Some of the positions available in marketing research include vice president of marketing research, research director, assistant director of research, project manager, field work director, statistician/data processing specialist, senior analyst, analyst, junior analyst and operational supervisor.

The most common entry-level position in marketing research for people with bachelor's degrees (e.g., BBA) is as operational supervisor. These people are responsible for supervising a well-defined set of operations, including field work, data editing, and coding, and may be involved in programming and data analysis.

Another entry-level position for BBAs is assistant project manager. An assistant project manager will learn and assist in questionnaire design, review field instructions, and monitor timing and costs of studies. In the marketing research industry, however, there is a growing preference for people with master's degrees. Those with MBA or equivalent degrees are likely to be employed as project managers.

A small number of business schools also offer a more specialized Master of Marketing Research (MMR) degree. An MMR typically prepares students for a wide range of research methodologies and focuses on learning both in the classroom and the field.

The typical entry-level position in a business firm would be junior research analyst (for BBAs) or research analyst (for MBAs or MMRs). The junior analyst and the research analyst learn about the particular industry and receive training from a senior staff member, usually the marketing research manager. The junior analyst position includes a training program to prepare individuals for the responsibilities of a research analyst, including coordinating with the marketing department and sales force to develop goals for product exposure.

The research analyst responsibilities include checking all data for accuracy, comparing and contrasting new research with established norms, and analysing primary and secondary data for the purpose of market forecasting.

As these job titles indicate, people with a variety of backgrounds and skills are needed in marketing research. Technical specialists such as statisticians obviously need strong backgrounds in statistics and data analysis. Other positions, such as research director, call for managing the work of others and require more general skills. To prepare for a career in marketing research, students usually:

- Take all the marketing courses.
- Take courses in statistics and quantitative methods.
- Acquire computer skills.

- Take courses in psychology and consumer behaviour.
- Acquire effective written and verbal communication skills.
- Think creatively.

Career ladder in marketing research:

1. Vice-President of Marketing Research: This is the senior position in marketing research. The VP is responsible for the entire marketing research operation of the company and serves on the top management team. Sets the objectives and goals of the marketing, research department.
2. Research Director: Also a senior position, the director has the overall responsibility for the development and execution of all the marketing research projects.
3. Assistant Director of Research: Serves as an administrative assistant to the director and supervises some of the other marketing research staff members.
4. (Senior) Project Manager: Has overall responsibility for design, implementation, and management of research projects.
5. Statistician/Data Processing Specialist: Serves as an expert on theory and application of statistical techniques. Responsibilities include experimental design, data processing, and analysis.
6. Senior Analyst: Participates in the development of projects and directs the operational execution of the assigned projects. Works closely with the analyst, junior analyst, and other personnel in developing the research design and data collection. Prepares the final report. The primary responsibility for meeting time and cost constraints rests with the senior analyst.
7. Analyst: Handles the details involved in executing the project. Designs and pretests the questionnaires and conducts a preliminary analysis of the data.
8. Junior Analyst: Handles routine assignments such as secondary data analysis, editing and coding of questionnaires, and simple statistical analysis.
9. Field Work Director: Responsible for the selection, training, supervision, and evaluation of interviewers and other field workers.

ANALYSIS AND MODELS IN MARKETING RESEARCH

Conjoint analysis is a statistical technique used in market research to determine how people value different features that make up an individual product or service. The objective of conjoint analysis is to determine what combination of a limited number of attributes is most influential on respondent choice or decision making. A controlled set of potential products or services is shown to respondents and by analysing how they make preferences between these products, the implicit valuation of the individual elements making up the product or service can be determined. These implicit valuations (utilities or part-worths) can be used to create market models that estimate market share, revenue and even profitability of new designs.

Conjoint originated in mathematical psychology and was developed by marketing professor Paul Green at the University of Pennsylvania and Data Chan. Other prominent conjoint analysis pioneers include professor V. "Seenu" Srinivasan of Stanford University who developed a linear programming (LINMAP) procedure for rank ordered data as well as a self-explicated approach, Richard Johnson (founder of Sawtooth Software) who developed the Adaptive Conjoint Analysis technique in the 1980s and Jordan Louviere (Ph.D., University of Iowa) who invented and developed Choice-based approaches to conjoint analysis and related techniques such as MaxDiff.

Today it is used in many of the social sciences and applied sciences including marketing, product management, and operations research. It is used frequently in testing customer acceptance of new product designs, in assessing the appeal of advertisements and in service design. It has been used in product positioning, but there are some who raise problems with this application of conjoint analysis.

Conjoint analysis techniques may also be referred to as multiattribute compositional modelling, discrete choice modelling, or stated preference research, and is part of a broader set of trade-off analysis tools used for systematic analysis of decisions.

These tools include Brand-Price Trade-Off, Simalto, and mathematical approaches such as evolutionary algorithms or Rule Developing Experimentation.

SUPPLY CHAIN MANAGEMENT

Supply chain management (SCM) is the management of a network of interconnected businesses involved in the ultimate provision of product and service packages required by end customers. Supply chain management spans all movement and storage of raw materials, work-in-process inventory, and finished goods from point of origin to point of consumption (supply chain).

Another definition is provided by the APICS Dictionary when it defines SCM as the "design, planning, execution, control, and monitoring of supply chain activities with the objective of creating net value, building a competitive infrastructure, leveraging worldwide logistics, synchronizing supply with demand and measuring performance globally."

Definitions

More common and accepted definitions of supply chain management are:

- Supply chain management is the systemic, strategic coordination of the traditional business functions and the tactics across these business functions within a particular company and across businesses within the supply chain, for the purposes of improving the long-term performance of the individual companies and the supply chain as a whole.
- A customer focused definition is given by Hines "Supply chain strategies require a total systems view of the linkages in the chain that work together efficiently to create customer satisfaction at the end point of delivery to the consumer. As a consequence costs must be lowered throughout the chain by driving out unnecessary costs and focusing attention on adding value. Throughput efficiency must be increased, bottlenecks removed and performance measurement must focus on total systems efficiency and equitable reward distribution to those in the supply chain adding value. The supply chain system must be responsive to customer requirements."
- Global supply chain forum - supply chain management is the integration of key business processes across the supply chain for the purpose of creating value for customers and stakeholders.
- The Council of Supply Chain Management Professionals (CSCMP), supply chain management encompasses the planning and management of all activities involved in sourcing, procurement, conversion, and logistics management. It also includes the crucial components of coordination and collaboration with channel partners, which can be suppliers, intermediaries, third-party service providers, and customers. In essence, supply chain management integrates supply and demand management within and across companies. More recently, the loosely coupled, self-organizing network of businesses that cooperate to provide product and service offerings has been called the Extended Enterprise.

A supply chain, as opposed to supply chain management, is a set of organizations directly linked by one or more of the upstream and downstream flows of products, services, finances, and information from a source to a customer. Managing a supply chain is 'supply chain management'.

Supply chain management software includes tools or modules used to execute supply chain transactions, manage supplier relationships and control associated business processes. Supply chain event management (abbreviated as SCEM) is a consideration of all possible events and factors that can disrupt a supply chain. With SCEM possible scenarios can be created and solutions devised.

Problems Addressed by Supply Chain Management

Supply chain management must address the following problems:

- *Distribution Network Configuration*: Number, location and network missions of suppliers, production facilities, distribution centers, warehouses, cross-docks and customers.
- *Distribution Strategy*: Questions of operating control (centralized, decentralized or shared); delivery scheme, *e.g.*,,, direct shipment, pool point shipping, cross docking, DSD (direct store delivery), closed loop shipping; mode of transportation, *e.g.*,,, motor carrier, including truckload, LTL, parcel; railroad; intermodal transport, including TOFC (trailer on flatcar) and COFC (container on flatcar); ocean freight; airfreight; replenishment strategy (*e.g.*,, pull, push or hybrid); and transportation control.
- *Trade-Offs in Logistical Activities*: The above activities must be well coordinated in order to achieve the lowest total logistics cost. Trade-offs may increase the total cost if only one of the activities is optimized. For example, full truckload (FTL) rates are more economical on a cost per pallet basis than less than truckload (LTL) shipments. If, however, a full truckload of a product is ordered to reduce transportation costs, there will be an increase in inventory holding costs which may increase total logistics costs. It is therefore imperative to take a systems approach when planning logistical activities. These trade-offs are key to developing the most efficient and effective Logistics and SCM strategy.
- *Information*: Integration of processes through the supply chain to share valuable information, including demand signals, forecasts, inventory, transportation, potential collaboration, etc.
- *Inventory Management*: Quantity and location of inventory, including raw materials, work-in-progress (WIP) and finished goods.
- *Cash-Flow*: Arranging the payment terms and methodologies for exchanging funds across entities within the supply chain.

Supply chain execution means managing and coordinating the movement of materials, information and funds across the supply chain. The flow is bi-directional.

Activities/Functions

Supply chain management is a cross-function approach including managing the movement of raw materials into an organization, certain aspects of the internal processing of materials into finished goods, and the movement of finished goods out of the organization and towards the end-consumer. As organizations strive to focus on core competencies and becoming more flexible, they reduce their ownership of raw materials sources and distribution channels. These functions are increasingly being outsourced to other entities that can perform the activities better or more cost effectively.

The effect is to increase the number of organizations involved in satisfying customer demand, while reducing management control of daily logistics operations. Less control and more supply chain partners led to the creation of supply chain management concepts. The purpose of supply chain management is to improve trust and collaboration among supply chain partners, thus improving inventory visibility and the velocity of inventory movement. Several models have been proposed for understanding the activities required to manage material movements across organizational and functional boundaries. SCOR is a supply chain management model promoted by the Supply Chain Council. Another model is the SCM Model proposed by the Global Supply Chain Forum (GSCF).

Supply chain activities can be grouped into strategic, tactical, and operational levels. The CSCMP has adopted The American Productivity and Quality Center (APQC) Process Classification Frameworka high-level, industry-neutral enterprise process model that allows organizations to see their business processes from a cross-industry viewpoint.

Strategic Level

- Strategic network optimization, including the number, location, and size of warehousing, distribution centers, and facilities.
- Strategic partnerships with suppliers, distributors, and customers, creating communication channels for critical information and operational improvements such as cross docking, direct shipping, and third-party logistics.
- Product life cycle management, so that new and existing products can be optimally integrated into the supply chain and capacity management activities.
- Information technology chain operations.
- Where-to-make and make-buy decisions.
- Aligning overall organizational strategy with supply strategy.
- It is for long term and needs resource commitment.

Tactical Level

- Sourcing contracts and other purchasing decisions.
- Production decisions, including contracting, scheduling, and planning process definition.
- Inventory decisions, including quantity, location, and quality of inventory.

- Transportation strategy, including frequency, routes, and contracting.
- Benchmarking of all operations against competitors and implementation of best practices throughout the enterprise.
- Milestone payments.
- Focus on customer demand and Habits.

Operational Level

- Daily production and distribution planning, including all nodes in the supply chain.
- Production scheduling for each manufacturing facility in the supply chain (minute by minute).
- Demand planning and forecasting, coordinating the demand forecast of all customers and sharing the forecast with all suppliers.
- Sourcing planning, including current inventory and forecast demand, in collaboration with all suppliers.
- Inbound operations, including transportation from suppliers and receiving inventory.
- Production operations, including the consumption of materials and flow of finished goods.
- Outbound operations, including all fulfillment activities, warehousing and transportation to customers.
- Order promising, accounting for all constraints in the supply chain, including all suppliers, manufacturing facilities, distribution centers, and other customers.
- From production level to supply level accounting all transit damage cases and arrange to settlement at customer level by maintaining company loss through insurance company.

Importance of Supply Chain Management

Organizations increasingly find that they must rely on effective supply chains, or networks, to compete in the global market and networked economy. In Peter Drucker's (1998) new management paradigms, this concept of business relationships extends beyond traditional enterprise boundaries and seeks to organize entire business processes throughout a value chain of multiple companies.

During the past decades, globalization, outsourcing and information technology have enabled many organizations, such as Dell and Hewlett Packard, to successfully operate solid collaborative supply networks in which each specialized business partner focuses on only a few key strategic activities. This inter-organizational supply network can be acknow-ledged as a new form of organization.

However, with the complicated interactions among the players, the network structure fits neither "market" nor "hierarchy" categories. It is not clear what kind of performance impacts different supply network structures could have on firms, and little is known about the coordination conditions and trade-offs that may exist among the players.

From a systems perspective, a complex network structure can be decomposed into individual component firms. Traditionally, companies in a supply network concentrate on the inputs and outputs of the processes, with little concern for the internal management working of other individual players. Therefore, the choice of an internal management control structure is known to impact local firm performance.

In the 21st century, changes in the business environment have contributed to the development of supply chain networks. First, as an outcome of globalization and the proliferation of multinational companies, joint ventures, strategic alliances and business partnerships, significant success factors were identified, complementing the earlier "Just-In-Time", "Lean Manufacturing" and "Agile Manufacturing" practices. Second, technological changes, particularly the dramatic fall in information communication costs, which are a significant component of transaction costs, have led to changes in coordination among the members of the supply chain network.

Many researchers have recognized these kinds of supply network structures as a new organization form, using terms such as "Keiretsu", "Extended Enterprise", "Virtual Corporation", "Global Production Network", and "Next Generation Manufacturing System". In general, such a structure can be defined as "a group of semi-independent organizations, each with their capabilities, which collaborate in ever-changing constellations to serve one or more markets in order to achieve some business goal specific to that collaboration".

The security management system for supply chains is described in ISO/IEC 28000 and ISO/IEC 28001 and related standards published jointly by ISO and IEC.

Historical Developments in Supply Chain Management

Six major movements can be observed in the evolution of supply chain management studies: Creation, Integration, and Globalization, Specialization Phases One and Two, and SCM 2.0.

- *Creation era*: The term supply chain management was first coined by a U.S. industry consultant in the early 1980s. However, the concept of a supply chain in management was of great importance long before, in the early 20th century, especially with the creation of the assembly line. The characteristics of this era of supply chain management include the need for large-scale changes, re-engineering, downsizing driven by cost reduction programmes, and widespread attention to the Japanese practice of management.

- *Integration era*: This era of supply chain management studies was highlighted with the development of Electronic Data Interchange (EDI) systems in the 1960s and developed through the 1990s by the introduction of Enterprise Resource Planning (ERP) systems. This era has continued to develop into the 21st century with the expansion of internet-based collaborative systems. This era of supply chain evolution is characterized by both increasing value-adding and cost reductions through integration. In fact a supply chain can be classified as a Stage 1, 2 or 3 network. In stage 1 type supply chain, various systems such as Make, Storage, Distribution, Material control, etc are not linked and are independent of each other. In a stage 2 supply chain, these are integrated under one plan and is ERP enabled. A stage 3 supply chain is one in which vertical integration with the suppliers in upstream direction and customers in downstream direction is achieved. An example of this kind of supply chain is Tesco.
- *Globalization era*: The third movement of supply chain management development, the globalization era, can be characterized by the attention given to global systems of supplier relationships and the expansion of supply chains over national boundaries and into other continents. Although the use of global sources in the supply chain of organizations can be traced back several decades (*e.g.,*, in the oil industry), it was not until the late 1980s that a considerable number of organizations started to integrate global sources into their core business. This era is characterized by the globalization of supply chain management in organizations with the goal of increasing their competitive advantage, value-adding, and reducing costs through global sourcing.
- *Specialization era—phase one*: Outsourced manufa-cturing and distribution: In the 1990s, industries began to focus on "core competencies" and adopted a specialization model. Companies abandoned vertical integration, sold off non-core operations, and outsourced those functions to other companies. This changed management requirements by extending the supply chain well beyond company walls and distributing management across specialized supply chain partnerships. This transition also re-focused the fundamental perspectives of each respective organization. OEMs became brand owners that needed deep visibility into their supply base. They had to control the entire supply chain from above instead of from within. Contract manufacturers had to manage bills of material with different

part numbering schemes from multiple OEMs and support customer requests for work -in-process visibility and vendor-managed inventory (VMI). The specialization model creates manufacturing and distribution networks composed of multiple, individual supply chains specific to products, suppliers, and customers who work together to design, manufacture, distribute, market, sell, and service a product. The set of partners may change just as to a given market, region, or channel, resulting in a proliferation of trading partner environments, each with its own unique characteristics and demands.

- *Specialization era—phase two*: supply chain manag-ement as a service: Specialization within the supply chain began in the 1980s with the inception of transportation brokerages, warehouse management, and non-asset-based carriers and has matured beyond transportation and logistics into aspects of supply planning, collaboration, execution and performance management. At any given moment, market forces could demand changes from suppliers, logistics providers, locations and customers, and from any number of these specialized participants as components of supply chain networks. This variability has significant effects on the supply chain infrastructure, from the foundation layers of establishing and managing the electronic communication between the trading partners to more complex requirements including the configuration of the processes and work flows that are essential to the management of the network itself. Supply chain specialization enables companies to improve their overall competencies in the same way that outsourced manufacturing and distribution has done; it allows them to focus on their core competencies and assemble networks of specific, best-in-class partners to contribute to the overall value chain itself, thereby increasing overall performance and efficiency. The ability to quickly obtain and deploy this domain-specific supply chain expertise without developing and maintaining an entirely unique and complex competency in house is the leading reason why supply chain specialization is gaining popularity. Outsourced technology hosting for supply chain solutions debuted in the late 1990s and has taken root primarily in transportation and collaboration categories. This has progressed from the Application Service Provider (ASP) model from approximately 1998 through 2003 to the On-Demand model from approximately 2003-2006 to the Software as a Service (SaaS) model currently in focus today.
- *Supply chain management 2.0*: Building on globalization and specialization, the term SCM 2.0 has been coined to describe both the

changes within the supply chain itself as well as the evolution of the processes, methods and tools that manage it in this new "era".

Web 2.0 is defined as a trend in the use of the World Wide Web that is meant to increase creativity, information sharing, and collaboration among users. At its core, the common attribute that Web 2.0 brings is to help navigate the vast amount of information available on the Web in order to find what is being sought. It is the notion of a usable pathway. SCM 2.0 follows this notion into supply chain operations. It is the pathway to SCM results, a combination of the processes, methodologies, tools and delivery options to guide companies to their results quickly as the complexity and speed of the supply chain increase due to the effects of global competition, rapid price fluctuations, surging oil prices, short product life cycles, expanded specialization, near-/far- and off-shoring, and talent scarcity. SCM 2.0 leverages proven solutions designed to rapidly deliver results with the agility to quickly manage future change for continuous flexibility, value and success. This is delivered through competency networks composed of best-of-breed supply chain domain expertise to understand which elements, both operationally and organizationally, are the critical few that deliver the results as well as through intimate understanding of how to manage these elements to achieve desired results. Finally, the solutions are delivered in a variety of options, such as no-touch via business process outsourcing, mid-touch via managed services and software as a service (SaaS), or high touch in the traditional software deployment model.

Supply Chain Business Process Integration

Successful SCM requires a change from managing individual functions to integrating activities into key supply chain processes. An example scenario: the purchasing department places orders as requirements become known. The marketing department, responding to customer demand, communicates with several distributors and retailers as it attempts to determine ways to satisfy this demand. Information shared between supply chain partners can only be fully leveraged through process integration.

Supply chain business process integration involves collaborative work between buyers and suppliers, joint product development, common systems and shared information. Lambert and Cooper (2000), operating an integrated supply chain requires a continuous information flow. However, in many companies, management has reached the conclusion that optimizing the product flows cannot be accomplished without implementing a process approach to the business.

The key supply chain processes stated by Lambert (2004) are:

- Customer relationship management

- Customer service management
- Demand management
- Order fulfillment
- Manufacturing flow management
- Supplier relationship management
- Product development and commercialization
- Returns management

Much has been written about demand management. Best-in-Class companies have similar characteristics, which include the following:

- Internal and external collaboration
- Lead time reduction initiatives
- Tighter feedback from customer and market demand
- Customer level forecasting

One could suggest other key critical supply business processes which combine these processes stated by Lambert such as:

- Customer service management
- Procurement
- Product development and commercialization
- Manufacturing flow management/support
- Physical distribution
- Outsourcing/partnerships
- Performance measurement

Customer Service Management Process

Customer Relationship Management concerns the relationship between the organization and its customers. Customer service is the source of customer information. It also provides the customer with real-time information on scheduling and product availability through interfaces with the company's production and distribution operations.

Successful organizations use the following steps to build customer relationships:

- Determine mutually satisfying goals for organization and customers
- Establish and maintain customer rapport
- Produce positive feelings in the organization and the customers

Bibliography

Aaker, D. A. : *Essentials of Marketing Research,* Wiley, New York, 2002.

Arnold, O.K. : *Applied Regression Analysis and Experimental Design*, London: CRC Press, 1985.

Assael, H. : *Consumer Behaviour and Marketing Action*, USA: PWS-Kent, 1992.

Baker, M. : *Marketing Management and Strategy*, Macmillan Business, 2000.

Beck, J.S. : *Medical Statistics on Microcomputers,* British Medical Journal, London, 1990.

Burnette, D.: *Teaching Qualitative Research: A Compendium of Model Syllabi*, Alexandria: Council on Social Work Education, 2003.

Chawla Dinesh *: Marketing Research*, Knowledge Book Dist, Delhi, 2010.

Chrzanowska, J. : *Interviewing Groups and Individuals in Qualitative Market Research,* Sage, London, 2002.

Churchill, G. A. : *Marketing research: methodological foundations,* South-Western/Thomson Learning, Mason, Ohio, 1995.

Cobb, G.W. : *An Electronic Companion to Business Statistics,* NY: Cognito Learning Media, Inc, 1997.

David, F.: *Quantitative Methods in Social Work: State of the Art,* New Delhi: Oxford University Press, 2001.

Finlay, B. : *Statistical Methods for the Social Sciences*, Dellen Publishing, Manas Dasgupta.

Fowler Jr., F. J. : *Survey Research Methods,* Newbury Park, London, 1993.

Girish Saxena: *Basic Concept of Media Economics: Theory and Practice*, Vista International Pub, Delhi, 2010.

Gordon, W. : *Goodthinking: A Guide to Qualitative Research,* Admap, Henley-on-Thames, 1999.

Hagle, T. : *Basic Math for Social Scientists - Problems and Solutions,* CA: Sage Publications, 1996.

Hague P., Hague N., Morgan C. : *Market Research In Practice,* Kogan Page, London, 2004.

Hague, P. N. : *Market Research: A Guide to Planning, Methodology and Evaluation,* Kogan Page, London, 2002.

Harrigan, K. R.: *Strategies for Declining Businesses*. Lexington, MA: Heath, 1980

Hunt, E. K. *History of Economic Thought, A Critical Perspective,* New York, HarperCollins, 1992.

Hussey, M. K. : *Quantitative Methods in Marketing,* International Thomson Business, London, 1999.

Isadore, Newman: *Qualitative-quantitative Research Methodology: Exploring,* Philippines: Oxford University Press, 2006.

Iversen, G.R. : *Bayesian Statistical Inference (Quantitative Applications in the Social Sciences),* Newbury Park, CA: Sage Publications, 1984.

Jackson, P. : *Buying Market Research,* Kogan Page, London, 1994.

Judy, L.: *Research for Effective Social Work Practice,* Philippines: Oxford University Press, 2006.

Kapila Mittal: *Research Methodology in Social Sciences,* Vista International, 2006.

Malhotra, N. K. : *Marketing Research: An Applied Orientation,* Prentice-Hall International, London, 2004.

Marczyk, G.R. : *Essentials of Research Design and Methodology,* Hoboken, NJ: John Wiley & Sons, 2005.

Monette, D.R.: *Applied Social Research: Tool for the Human Services,* New York: Holt, 2002.

Moore, D. S. : *Statistics: Concepts and Controversies,* W. H. Freeman, New York, 2001.

Morgenstern, H. : *Epidemiologic Research: Principles and Quantitative Methods,* Belmont, CA : Lifetime Learning Publications, 1982.

Proctor, T. : *Essentials of Marketing Research,* Financial Times Prentice Hall, Harlow, 2003.

Reid, W.J.: *Research in Social Work,* New York: Columbia University Press, 2006.

Russell, K.: *Fundamentals of Social Work Research,* New Delhi: Oxford University Press, 2001.

Saunders, M., Lewis, P. : *Research Methods for Business Students,* Financial Times Prentice Hall, Harlow, 2003.

Secic, M. : *How to Report Statistics in Medicine,* PA:American College of Physicians, 1997.

Sleight, P. : *Targeting Customers: How to Use Geodemographic and Lifestyle Data in Your Business,* World Advertising

Vogt, W.P. : *Dictionary of Statistics and Methodology: A Nontechnical Guide for the Social Sciences,* CA: Sage Publications, 2005.

Webb, J. R. : *Understanding and Designing Market Research,* Thomson Learning, London, 2002.

Weber, M.: *The Theory of Social and Economic Organizations,* New York, Oxford University Press, 1947.

Wichern, D. W. : *Business Statistics: Decision Making with Data,* Wiley, Chichester, 1997.

Wilson, A. : *Industrial Marketing Research: Management and Technique,* Huchison, London, 1979.

Wright, L. T. : *The Marketing Research Process,* Prentice Hall, London. 1995, 2001.

Index